Past Revolutions, Future Transformations

What can the history of revolutions in military affairs tell us about transforming the U.S. military?

RICHARD O. HUNDLEY

Prepared for the
Defense Advanced Research Projects Agency

National Defense Research Institute

RAND

The research described in this report was sponsored by the Defense Advanced Research Projects Agency. The research was conducted in RAND's National Defense Research Institute, a federally funded research and development center supported by the Office of the Secretary of Defense, the Joint Staff, the unified commands, and the defense agencies, Contract DASW01-95-C-0059.

Library of Congress Cataloging-in-Publication Data

Hundley, Richard O.
Past revolutions, future transformations : what can the history of revolutions in military affairs tell us about transforming the U.S. military? / Richard O. Hundley.
p. cm.
"Prepared for the Defense Advanced Research Project Agency (DARPA) by RAND's National Defense Research Institute."
"MR-1029-DARPA."
Includes bibliographical references (p.).
ISBN 0-8330-2709-3
1. Military art and science. 2. Military history. 3. United States—Armed Forces—Organization. I. Title.
U104.H89 1999
355.4 ' 0973—dc21 99-25786
CIP

Published 1999 by RAND
1700 Main Street, P.O. Box 2138, Santa Monica, CA 90407-2138
1333 H St., N.W., Washington, D.C. 20005-4707
RAND URL: http://www.rand.org/
To order RAND documents or to obtain additional information, contact Distribution Services: Telephone: (310) 451-7002; Fax: (310) 451-6915; Internet: order@rand.org

PREFACE

The history of the 20th century has shown that advances in technology can bring about dramatic changes in military operations, often termed "revolutions in military affairs" or RMAs. Technology-driven RMAs have been occurring since the dawn of history, they will continue to occur in the future, and they will continue to bestow a military advantage on the first nation to develop and use them. Accordingly, it is important to the vitality and robustness of the U.S. defense posture for the Department of Defense (DoD) research and development (R&D) community to be aware of technology developments that could revolutionize military operations, and for the U.S. military services to be on the lookout for revolutionary ways in which to employ those technologies in warfare.

This leads to three interrelated questions:

- Regarding past revolutions in military affairs (RMAs),
  - What lessons can we learn from the historical record?
- Regarding being prepared for future RMAs carried out by others,
  - What does it take to be prepared?
- Regarding transforming U.S. military forces by carrying out our own RMAs,
  - What does it take to be successful?

This report addresses these questions, which are particularly relevant today when the DoD has set out on a concerted effort to bring about a technology-driven transformation of the U.S. military to achieve the operational goals outlined in *Joint Vision 2010* (Shalikashvili, 1996).

This research was sponsored by the Director's Office of the Defense Advanced Research Projects Agency (DARPA) and conducted by the Acquisition and Technology Policy Center of RAND's National Defense Research Institute (NDRI). NDRI is a federally funded research and development center sponsored by the Office of the Secretary of Defense, the Joint Staff, the defense agencies, and the unified commands.

CONTENTS

FIGURES

TABLES

SUMMARY

As the Gulf War showed, advances in technology can bring about dramatic changes in military operations. Such technology-driven changes in military operations will continue to bestow a military advantage on the first nation to develop and use them. Accordingly, the vitality and robustness of the U.S. defense posture depend on the DoD R&D community being on the leading edge of breakthrough technologies that could revolutionize military operations. Also, the U.S. military services must be on the lookout for revolutionary ways in which to employ those technologies in warfare.

LESSONS FROM PAST RMAs

The Characteristics of Revolutions in Military Affairs

Based on an examination of two historical records—the long history of military technology and the military "revolutions" in the 20th century—we conclude that the *defining characteristic* of a revolution in military affairs (RMA) can be stated as follows:

> An RMA involves a paradigm shift in the nature and conduct of military operations
>
> - which either *renders obsolete or irrelevant* one or more *core competencies* of a dominant player,
> - or creates one or more new core competencies, in some new dimension of warfare,
> - or both.

We elaborate on the key terms in this definition in Chapter Two.

Based once again on the historical record, we can list other notable characteristics of RMAs:

- RMAs are rarely brought about by dominant players.
- RMAs frequently bestow an enormous and immediate military advantage on the first nation to exploit them in combat.
- RMAs are often adopted and fully exploited first by someone other than the nation inventing the new technology.
- RMAs are not always technology-driven.
- Technology-driven RMAs are usually brought about by combinations of technologies, rather than individual technologies.
- Not all technology-driven RMAs involve weapons.
- All successful technology-driven RMAs appear to have three components: technology, doctrine, and organization.
- There are probably as many failed RMAs as successful RMAs.
- RMAs often take a long time to come to fruition.
- The military utility of an RMA is frequently controversial and in doubt up until the moment it is proven in battle.

We provide historical examples of each of these characteristics in Chapter Two.

Paradigm shifts are not limited to the military arena. They occur in the business world as well, where in recent years they have been a much studied phenomenon. One of the clear messages from the business literature is that paradigm shifts that destroy core business competencies are rarely brought about by dominant players. (This reinforces one of the RMA characteristics noted above.)

The Nature of the Breakthrough Process Leading to RMAs

RMAs are the result of multiple innovations:

- *A new technology* (or several new technologies), which enables devices and systems not previously possible or contemplated.

- *A new device,* based on this new technology, which does something not previously doable.
- *A new system,* based on this new device, which performs a military function either dramatically better or dramatically differently than it had been performed before, or performs a new military function.
- *A new operational concept,* which describes the manner in which the new system is employed in some type of military situation, accomplishing some military task either dramatically better or dramatically differently than it had been accomplished before, or performing a new task that did not exist previously.
- *A new doctrine and force structure*—doctrine that codifies the principles governing the employment of the new system and force structure that provides the military organization necessary to fully realize its potential.

These various stages culminate in *a new military reality,* in which a paradigm shift has occurred in some segment of the military arena.

There are several important features of this breakthrough process leading to RMAs:

- Unmet military challenges are an essential element driving creativity at each step in the process. Without one or more existing challenges, technologies are unlikely to be combined into devices and devices into systems, and new operational concepts, doctrine, and force structures are unlikely to be developed.
- The various innovations sometimes occur out of order: e.g., an operational concept is "invented" before a technology, device, and/or system exists adequate to realize its postulated potential.
- RMAs can fail in the face of obstacles at any step in the chain. The necessary technology may exist but the contemplated devices prove impractical. It may not be possible to turn the new devices into viable systems. No operational concept may exist to employ an otherwise viable system concept. The force structure necessary to exploit the operational concept may not exist because the operational concept is unacceptable to the prevailing military culture, or because the new force structure requires too

large a change in existing military organizations. (We provide historical examples of each of these situations in Chapter Three.)

BEING PREPARED FOR FUTURE RMAs (CARRIED OUT BY OTHERS)

Being Aware of the Next RMA: The Observables of the Emergence of New RMAs

The first step in being prepared for future RMAs carried out by others is *being aware* that an RMA may be occurring. Much of the RMA process can be observed and anticipated, as discussed in Chapter Four. This is particularly true during the exploitation and selling phase that leads from a new device, operational concept, and system concept to a new doctrine and force structure—and which is absolutely essential if the new doctrine and force structure required to truly realize an RMA are ever to be achieved. During these phases of the RMA process, readily observable signals are produced in a number of venues, including:

- *Various press organs,* including the trade press (defense, aerospace, etc.), the military art and science press, the science and technology press, the international security and foreign affairs press, and the general business press, as well as leading newspapers and magazines.
- *The worldwide arms market,* both legitimate and clandestine.
- *Inferior military establishments,* which are trying to leapfrog the dominant military players.
- *Dominant military players,* who are trying to discredit new ideas that threaten their core competencies.
- *Military research, development, test, and evaluation (RDT&E) activities,* particularly those involving new technologies, systems, and/or operational concepts.

Some of these venues are open, some are closed. The activities in open venues are usually readily observable by almost anyone; the activities in closed venues are normally shut off from view by outsiders. Open and closed venues obviously require different information-collection approaches.

Collection Is Not Enough; Assessment Is Also Required

Not all potential RMAs come to pass; many are aborted and fall by the wayside, for a variety of reasons. Accordingly, the collection of observables to the emergence of new RMAs is not enough; these observables must also be carefully assessed, to separate out the serious RMA candidates from all of the wild-eyed dreams. A multistep collection and assessment process is required, with the following components:

- *An initial, wide-area-search collection process,* to detect any and all RMA visions and dreams, no matter where they arise throughout the world, no matter how far out they may appear. The output of this continually ongoing collection activity is a living list of RMA "visions and dreams."
- *An initial screening process,* based on some sort of plausibility criteria, to weed out the "antigravity" ideas[1] (or their equivalent) from this list but keep in all those items with some prospects of success. The output of this step is a list of potential RMA candidates.
- *A monitoring collection process,* focused on each of these potential RMA candidates and continuing over an extended period.
- *A more careful assessment process,* which could include challenges, hurdles, and tests that a candidate RMA must pass. The output of this ongoing step is a list of serious RMA candidates, to be closely monitored and reassessed as they evolve and mature.

This process requires patience and staying power. Since future RMAs cannot be scheduled, one must establish a collection and assessment process that can endure over a long time.

---

[1]By antigravity ideas we mean concepts that are clearly not feasible, based on fundamental physical or engineering considerations.

The Essential Elements of a Worldwide RMA Breakthrough Watch and Assessment Activity

This leads us to the essential elements of a worldwide RMA breakthrough watch and assessment activity:

- An *information collection activity* that conducts two types of collection: *worldwide search,* primarily open source, to uncover new RMA visions, and *continued monitoring,* using open source and (if necessary) standard closed-source techniques, and focused on specific RMA candidates that have survived the initial screening process.
- An *RMA assessment activity* that conducts two types of assessment: *initial screening,* which keeps in all those items with some prospects of success, and *continued and more careful assessment,* over time, to follow potential/serious RMA candidates as they mature to see if they surmount various challenges and hurdles.

These collection and assessment activities can be carried out in two separate but closely coupled organizations—information collection in some sort of intelligence organization and RMA assessment in some sort of an advanced military research and development organization, or they can be carried out in one organization having combined capabilities. Whichever way it is done, such a worldwide RMA breakthrough watch and assessment activity should ensure U.S. awareness of future RMAs being carried out by others, if properly implemented in an enduring fashion.[2]

Being Responsive to an Emerging RMA Is a More Difficult Challenge

Being aware of emerging RMAs is not enough; one must also be responsive. History is full of examples of military organizations that were aware of an emerging RMA but failed to respond in an adequate

---

[2]It is vitally important that this RMA breakthrough watch and assessment activity *endure* over long periods, since one cannot predict when an RMA harmful to U.S. military capabilities may arise. The current U.S. focus on "the RMA" may constitute an informal, temporary breakthrough watch. We are proposing a formal, more permanent one.

fashion. Failure to respond can lead a nation to military disaster just as easily as unawareness can.

History shows that established military organizations more often than not fail to respond adequately to emerging RMAs threatening their core competencies, even RMAs of which they are aware. This occurs primarily because of inherent obstacles to the changes necessary to cope with an RMA. These obstacles are not unique to military organizations; rather they are for the most part generic, psychological obstacles to the organizational learning and change necessary to cope with paradigm shifts threatening core competencies, no matter what their shape or form.

In recent years, analysts in both the military and business arenas have addressed this problem, characterizing the various obstacles to organizational learning and change in the face of paradigm shifts, and identifying proven techniques to overcome them. Using this literature as our point of departure, we have identified the following characteristics of what we would term a future-oriented military organization likely to respond adequately to an emerging RMA:

- "Productive paranoia"[3] regarding the future.
- A continually refined vision of how war may change in the future.
- An organizational climate encouraging vigorous debate regarding the future of the organization.
- Mechanisms available within the organization for experimentation with new ideas, even ones that threaten the organization's current core competencies.
- Senior officers with traditional credentials willing to sponsor new ways of doing things.
- New promotion pathways for junior officers practicing a new way of war.

---

[3]We have coined this term to capture the major theme expressed by Andrew Grove, the former CEO of Intel, in his recent book regarding paradigm shifts in the business world, *Only the Paranoid Survive.*

We expand on each of these characteristics in Chapter Five.

Possessing these characteristics is no guarantee of future success. However, a military establishment lacking one or more of them is less likely to respond adequately to an emerging RMA being carried out by others.

BRINGING ABOUT FUTURE RMAs (OF YOUR OWN)

What about developing your own RMA, rather than merely responding to someone else's? History suggests that for a military organization to bring about an RMA of its own all of the following items are probably necessary:

- You must have a fertile set of enabling technologies.[4]
- You must have unmet military challenges.
- You must focus on a definite "thing" or a short list of "things"—a device or system exploiting the enabling technologies together with a concept for its operational employment.[5]
- You must ultimately challenge someone's core competency.
- You must have a receptive organizational climate, which fosters a continually refined vision of how war may change in the future and which encourages vigorous debate regarding the future of the organization.
- You must have support from the top: senior officers with traditional credentials willing to sponsor new ways of doing things and able to establish new promotion pathways for junior officers practicing a new way of war.
- You must have mechanisms for experimentation, to discover, learn, test, and demonstrate.[6]

---

[4]Assuming we are talking about a technology-driven RMA.

[5]This focusing process can take considerable time; until it occurs there is no RMA.

[6]The purpose of these experiments is to *discover* what you can do militarily with new technologies and combinations of new technologies; to *learn* which combinations of devices, systems, and operational concepts work better and which do not work as well; to *test* promising devices, systems, and operational concepts in a wide variety of real-

- You must have some way of responding positively to the results of successful experiments, in terms of doctrinal changes, acquisition programs, and force structure modifications.

We elaborate on each of these in Chapter Six .

With all of these things—and at least one brilliant idea—a military organization has a reasonable chance of bringing about a successful RMA. Without any one of these elements, the chances are much less, even if there is a brilliant idea, and history suggests the RMA process is likely to fail.

Today's Force Transformation/RMA Activities

Since publication of the 1997 *Quadrennial Defense Review* (QDR) (Cohen, 1997), the DoD has been involved in a concerted effort to "transform" the U.S. military, motivated by a fourfold set of objectives:

- to achieve the operational goals outlined in *Joint Vision 2010* (JV2010) (dominant maneuver, precision engagement, full-dimensional protection, focused logistics),
- to bring about the cost savings necessary to pay for force modernization,
- to achieve a new, affordable force structure that can be maintained in the future, and
- to take advantage of the [so-called] revolution in military affairs currently ongoing—"the RMA."[7]

---

world circumstances, thereby focusing on the combination of device(s), system(s), and employment concept(s) most likely to bring about an RMA; and finally to *demonstrate* that the chosen set of device(s), system(s), and operational concept(s) offers the potential for a revolutionary improvement in military capabilities in real-world conflict situations.

[7]Based on the definition of an RMA used here—a paradigm shift upsetting a core competency of a dominant player or creating a new core competency in some new dimension of warfare—it is too early to tell if the current military-technical revolution will result in one or more true RMAs. The jury is still out.

DoD force transformation activities under way thus far include the development of several future visions of warfare, the establishment of a number of battle laboratories and warfighting centers dedicated to exploring new ways of warfare, a number of wargames exploring new ways of warfare, a number of developmental and field experiments, and some new organizational arrangements. These various activities are pursuing a large number of technology/device/system/operational employment concept combinations, many of which probably represent evolutionary improvements on current ways of waging war, but several of which could possibly lead to RMAs. Several different concepts have been proposed as the kernel of "the RMA," including long-range precision fires, information warfare, a "system of systems," "network centric warfare," and a "cooperative engagement capability."

Does anything appear to be missing in these current DoD force transformation/RMA activities? Based on the history of past RMAs and the RMA checklist above, the answer seems to be "yes." Table S.1 summarizes our assessment; we elaborate on this in Chapter Seven.

"The RMA": Where We Seem to Be Today

Using Secretary Cohen's QDR terminology to describe the force transformation process, where is "the RMA" today? Harking back to the model of the RMA process we presented earlier, we can say the following:

- *New technology.* We have a lot of this.
- *New devices and systems.* We have a lot of ideas for new devices and systems. Many (but not all) of them have been or are being built. Some (but not most) of them are undergoing experiments, but not necessarily risky experiments, covering the entire discover, learn, test, and demonstrate spectrum.
- *New operational concepts.* We have many of these, each with their advocates and detractors. A few are undergoing actual experiments. Most are still in paper discussions and arguments.
- *New doctrine and force structure.* We are a long way from this, a very long way.

Table S.1

Does Anything Appear to Be Missing from DoD's Current RMA Activities?

RMA Checklist	DoD's Current Situation
You must have a fertile set of enabling technologies	*We clearly have this*
You must have unmet military challenges	*We have several of these (but are they compelling enough?)*
You must have a receptive organizational climate	*We may have this in some Services (but not in others)*
You must have support from the top	*We have this (but does it include all of the Services?)*
You must have mechanisms for experimentation (to discover, learn, test, and demonstrate)	*We have these (but do they cover the entire discover, learn, test, and demonstrate spectrum, and do they encourage "risky" experiments?)*
You must focus on a definite "thing" or a short list of "things"	*Thus far, this seems to be missing*
You must ultimately challenge someone's core competency	*Thus far, this seems to be missing*
You must have ways of responding positively to successful experiments (in terms of doctrine, acquisition, and force structure)	*This could be a problem (can the DoD system respond positively to a risky new idea involving radical change?)*

We are also a long way from focusing on a short list of potentially revolutionary devices, systems, and operational concepts around which we can transform the force. This necessary focusing process could take a few years, probably will take several years, and possibly will take many years.[8]

Another concern: In most past RMAs, the force was not transformed—i.e., old force structure elements replaced by RMA elements—until the RMA had been proven in battle. Until then, the RMA elements were treated as add-ons to the then-existing force structure. Based on the QDR, the DoD appears to be planning to

[8]One or more true RMAs, in the sense defined here, are probably required to transform the force to the extent postulated in the QDR: a lot more capability for a lot fewer resources.

transform the force, i.e., replacing old elements with new RMA elements rather than merely adding those elements, before "the RMA" is proven in combat. This flies in the face of history.

What Needs to Be Done?

Based on the history of past RMAs, there appear to be some key elements missing in DoD's current force transformation activities:

- None of the Services' current core competencies are being challenged;
- There is inadequate focus on a definite "thing" or a short list of "things";
- The DoD acquisition system may not be adequately receptive to novel/radical innovations.

These missing elements can be filled in by:

- Setting up DoD concept groups and experimental groups to identify and experiment with new systems and operational concepts that (a) challenge current Service core competencies and (b) increase the focus of the current RMA efforts;
- Establishing provisional operational units to participate in experiments with new systems and operational concepts;
- Establishing a new branch in the DoD acquisition system that tolerates substantial uncertainties regarding military utility to a much later stage in the acquisition process.

We elaborate on each of these in Chapter Seven.

Doing the above will facilitate DoD's force transformation activities and help ensure that the next RMA is brought about by the United States and not some other nation.

ACKNOWLEDGMENTS

The intellectual foundations for the material presented in this report, along with many of the details, were originally developed in a series of 1995 working group sessions involving the author and four of his RAND colleagues: Bruno W. Augenstein, Steven C. Bankes, James A. Dewar, and Samuel Gardiner. Without their contributions, the author could not have written this report.

The research also benefited greatly from subsequent discussions with Samuel Gardiner on revolutions in both military and business affairs; Paul K. Davis and Eugene C. Gritton on DoD's current force transformation process; John Birkler and Giles Smith on changes in the DoD acquisition process; Martin Libicki on the recent RMA-related literature; and F. L. (Frank) Fernandez, the Director of DARPA, and Thomas Tesch, Office of the Undersecretary of the Navy, regarding innovation in military and business affairs. In addition, Paul Davis and Martin Libicki reviewed an early draft of this report and contributed several insightful comments.

The author is deeply grateful to all of these individuals for their numerous suggestions and insights.

Chapter One

INTRODUCTION

As illustrated by the Gulf War, recent advances in technology have brought about dramatic changes in military operations: the use of low-observable aircraft to negate air defenses, smart weapons for precision conventional-strike operations, the employment of both ballistic missiles and antiballistic missiles (ABMs) in conventional warfare, and so forth. These dramatic technology-driven changes in military operations, sometimes termed a revolution in military affairs (RMA), are not unique in the history of warfare, but merely the latest in a chain of breakthrough technologies[1] extending back over time and including examples such as the ironclad in the 1860s, the machine gun in the 1890s–1910s, the manned aircraft and the tank in the 1920s–1930s, the aircraft carrier and radar in the 1930s–1940s, and nuclear weapons in the 1940s–1950s.[2]

Such technology-driven breakthroughs in military operations will continue to occur, and they will continue to bestow a military advantage on the first nation to develop and use them. Accordingly, the Department of Defense (DoD) research and development (R&D) community must be on the leading edge of breakthrough technolo-

[1]The term breakthrough technologies was first used (in recent times) by the Defense Science Board and Director Defense Research and Engineering (DDR&E) in 1990–1991 in conjunction with major technology-driven shifts in the nature and conduct of military operations. (See DSB, 1990, and Herzfeld, 1991.)

[2]This list includes just some of the more recent examples. The longbow, developed by the English during the 13th century and used against the French with devastating effect at Crecy (1346), Poitiers (1356), and Agincourt (1415) during the Hundred Years' War, is an earlier example of a breakthrough technology in the military arena. See Churchill (1958), pp. 332–351, 354–357, and 400–408.

gies that could revolutionize military operations in the future, and the U.S. military services must be on the lookout for revolutionary ways in which to employ those technologies in warfare.

This leads to three interrelated questions:

- Regarding past revolutions in military affairs,
  - What lessons can we learn from the historical record regarding the characteristics of RMAs and of the breakthrough process leading to RMAs?
- Regarding being prepared for future RMAs carried out by others,
  - What will it take for the United States to anticipate and be prepared for future technology-driven RMAs carried out by others?
- Regarding transforming U.S. military forces by carrying out our own RMAs,
  - What does it take to be successful?

This report addresses these three questions, which are particularly relevant today when the DoD has set out on a concerted effort to bring about a technology-driven transformation of the U.S. military to achieve the operational goals outlined in *Joint Vision 2010*.[3]

Regarding the first of our three topics—lessons to be learned from past RMAs—we begin in Chapter Two by identifying and describing a number of significant characteristics of RMAs and discussing the relationship between breakthrough technologies and RMAs. In Chapter Three we develop a number of models describing various aspects of the breakthrough process leading to RMAs. In both of these chapters we have taken as our point of departure the historical record of past technology-driven revolutions, in both military affairs and in the business world. We have also considered the specific

---

[3]See Shalikashvili (1996) for a discussion of *Joint Vision 2010*. See the *Quadrennial Defense Review* (Cohen, 1997) for a high-level statement of DoD's plans to bring about a technology-driven transformation of the U.S. military.

lessons learned in RAND's recent investigation of RMAs for the Office of Net Assessment.[4]

Regarding our second topic—being prepared for future RMAs carried out by others—in Chapter Four we identify a number of observables that could be used to anticipate the emergence of new RMAs. We use these observables as the foundations for a worldwide RMA breakthrough watch and assessment activity, which could be used to monitor and assess worldwide developments in technology and operational military concepts that might give rise to future RMAs.

As the historical record shows, being aware of an emerging RMA is not enough to avert military disaster; a nation must also be responsive to the implications of that RMA. This can be a difficult challenge, particularly for a dominant military player such as the United States is today. We discuss this second challenge in Chapter Five, where we identify the characteristics of a future-oriented military organization likely to respond adequately to an emerging RMA.

Regarding our final topic—successfully carrying out one's own RMAs—in Chapter Six we list the various elements that history suggests are necessary to bring about a successful RMA. In Chapter Seven, we compare DoD's current force transformation activities with this list and ask: Is anything missing to bring about an RMA? The answer, in our view, appears to be yes. We conclude by discussing what can be done to fill in these (seemingly) missing elements.

---

[4]In 1995, Sam Gardiner and Daniel Fox of RAND carried out an extensive series of wargaming exercises to investigate the RMA process.

PART I. LESSONS FROM PAST RMAs

Chapter Two

THE CHARACTERISTICS OF REVOLUTIONS IN MILITARY AFFAIRS

Technology-driven changes in military operations are not recent phenomena. Indeed, technological developments have been bringing about profound changes in the nature of warfare since the dawn of history.[1] Brodie (1973), Dupuy (1984), and van Creveld (1989) provide overviews of the historical panorama of military technology and its impact on warfare over the last (roughly) 4000 years, from the earliest developments (e.g., the chariot) to the most recent (e.g., nuclear weapons).

Beginning with the Soviet focus on the so-called military-technical revolution[2] and continuing with the work initiated by the Office of Net Assessment on the current revolution in military affairs,[3] considerable attention has been paid to the sometimes revolutionary nature of advances in military technology, with particular focus on events in

---

[1]A rich literature of the history of military technology describes this process. Van Creveld (1989) includes a bibliographical essay reviewing this literature, with numerous references.

[2]In 1984, Marshal Nikolai V. Ogarkov and other Soviet military thinkers began to stress that the emergence of advanced, nonnuclear technologies was engendering a new military-technical revolution in military affairs. See FitzGerald (1987) for an overview of Soviet thought on this subject.

[3]See Marshall (1993 and 1995) for original statements of the views of Andrew Marshall, the Director of the Office of the Secretary of Defense (OSD) Office of Net Assessment, regarding the current revolution in military affairs. Ricks (1994) contains an early published discussion of these views.

the 20th century. This has led to a resurgence of writing on the subject.[4]

We use these two historical records—of the long sweep of military technology and of the military revolutions in the 20th century—as our point of departure in describing the characteristics of RMAs.

WHAT IS AN RMA?

Much has been written recently regarding the current RMA, which is often viewed as

> a military technical revolution combining [technical advances in] surveillance, C3I [command, control, communications, and intelligence] and precision munitions [with new] operational concepts, including information warfare, continuous and rapid joint operations (faster than the adversary), and holding the entire theater at risk (i.e., no sanctuary for the enemy, even deep in his own battlespace).[5]

A number of people have written regarding this RMA, including Kendall (1992), Marshall (1993 and 1995), Mazarr et al. (1993), Mazarr (1994), Krepinevich (1994 and 1995), Libicki and Hazlett (1994), Gray (1995), Barnett (1996), Libicki (1996 and 1999), Blaker (1997), Buchan (1998), and Davis et al. (1998).[6] This literature well describes the elements of the current RMA, but does not shed much light on the characteristics of RMAs in general. That is, it does not address questions such as: How does one describe generically what constitutes an RMA? What are the defining characteristics of an RMA? To answer these questions, we must turn to the historical record of technology-driven changes in military operations.

Based on the historical record, it appears that the defining characteristic of an RMA can be stated as follows:

---

[4]See Krepinevich (1994), Murray and Watts (1995), Gray (1995), Bartlett et al. (1996), Libicki (1996), Murray and Millet (1996), and Blaker (1997) for a sampling of this recent literature.

[5]See McKendree (1996).

[6]The various DoD science boards have also discussed the current RMA, not always by name. For example, see SAB (1995), DSB (1996), and NSB (1997).

An RMA involves a paradigm shift in the nature and conduct of military operations

- which either *renders obsolete or irrelevant* one or more *core competencies* of a dominant player,
- or creates one or more new core competencies, in some new dimension of warfare,
- or both.

There are a number of key terms in this definition:

- *Paradigm.* An accepted model that serves as the basic pattern for a segment of military operations.[7] For example, opposing infantry units arranged in orderly formations maneuvering in the open to engage each other at close quarters, with supporting artillery fire, was the operational paradigm for land combat during the Napoleonic Wars. Opposing warships arranged in line-of-battle on parallel courses and engaging with gunfire was the operational paradigm for naval fleet engagements during those same wars, as well as during the First World War 100 years later.
- *Core competency.* A fundamental ability that provides the foundation for a set of military capabilities. For example, the ability to detect vehicular targets from the air and attack them with precision weapons is today a core competency of the U.S. Air Force. In the period between World War I and II, the ability to deliver accurate naval gunfire at ranges upwards of 20 miles was a core competency of the surface combat units of the U.S. Navy. In the 13th and 14th centuries, the ability of a longbowman to put an arrow accurately through the chain mail armor of a knight on horseback or a man-at-arms on the ground at ranges of 250–300 yards was a core competency of the English archers.[8]

---

[7]Paradigms also play a central role in other areas of human endeavor. For example, Kuhn (1970) discusses the role of paradigms and paradigm shifts in science. Likewise, Barker (1992) and Grove (1996) discuss the role of paradigms and paradigm shifts in business. Grove uses the term "strategic inflection point" rather than "paradigm shift" to denote the phenomenon, but the meaning is the same.

[8]Dupuy (1984, pp. 81–84) and Burke (1978, pp. 59–62) discuss the capabilities of the English longbowmen.

- *Dominant player.* A military organization that possesses a dominating set of capabilities in an area of military operations. For example, today the U.S. Air Force is the dominant player in air-to-air combat and air-to-ground attack. At the end of World War II, the carrier force of the U.S. Navy was the dominant player in naval warfare. At the end of World War I, the battle fleets (e.g., the battleship and battle cruiser forces) of the British Navy and the U.S. Navy were the dominant players in naval surface warfare. Going back further in history, during the Middle Ages the armored cavalry (i.e., knights on horseback) was the dominant player in land warfare in Europe. Even further back, in Roman times the Roman legion was the dominant player in land warfare throughout the Roman Empire.
- *Dimension of warfare.* The dimension on which warfare is conducted, the first and most ancient of which was the land surface of the earth (land warfare). The second and almost as ancient dimension on which warfare was conducted was the water surface of the earth (naval warfare). In the 20th century several new dimensions were added: the underwater portions of the oceans (undersea warfare), the air above the earth's surface (air warfare), and the homelands of the combatants (strategic warfare and intercontinental warfare). Another dimension much talked about since the Second World War but in which actual combat has not yet occurred is the region outside the earth's atmosphere (space warfare). As the information revolution continues, there is increasing discussion of cyberspace as still another dimension of warfare (information warfare).[9]
- *Paradigm shift.* A profound change in the fundamental model underlying a segment of military operations. For example, the carrier warfare paradigm, in which opposing naval forces engaged each other at 100- to 200-mile distances without ever coming within naval gunfire range, represented a profound change in the basic model underlying naval warfare. It rendered obsolete the core (naval gunfire) competency of the hitherto

---

[9]See Toffler (1993), Molander et al. (1996), and Arquilla and Ronfeldt (1997) for three views of this newest dimension of warfare.

dominant battleship fleets, and was therefore an RMA.[10] The blitzkrieg paradigm, in which highly mobile armored forces broke through enemy lines and rapidly penetrated to the rear, represented a profound change in the basic model underlying land warfare. It rendered obsolete the core competency of the hitherto dominant infantry and artillery forces for static defenses of prepared positions, and was therefore an RMA.[11] The nuclear-warhead-tipped intercontinental ballistic missile (ICBM) created a new core competency (an overwhelming, virtually unstoppable ability to destroy cities and other large-scale targets in the homeland of an opponent thousands of miles away) in a new dimension of warfare (intercontinental strategic warfare), and was therefore an RMA.

If a development in military technology does not either render obsolete a core competency of a dominant player or create a new core competency, it is not an RMA. If it does, it is.[12] Table 2.1 gives a few illustrative examples of developments in military technology that satisfy this criteria.

OTHER NOTABLE CHARACTERISTICS OF RMAs

Based on the historical record, other notable characteristics of RMAs are:

- *RMAs are rarely brought about by dominant players.* For example, during the period between the First and Second World Wars, the French and British infantry and artillery forces, the dominant

---

[10]The Battle of the Coral Sea (1942) was the first engagement in which this new paradigm played a dominating role. See Morison (1963), pp. 140–147.

[11]There are many descriptions of the development and impact of the blitzkrieg paradigm. Guderian (1952) provides a subjective, firsthand view; Corum (1992) provides a more objective, balanced presentation.

[12]Krepinevich (1994) has proposed a logically similar definition of an RMA: "What is a military revolution? It is what occurs when the application of new technologies into a significant number of military systems combines with innovative operational concepts and organizational adaptation in a way that fundamentally alters the character and conduct of conflict." We prefer our wording because of the emphasis it places on *changes in core competencies* as central to the RMA process.

Table 2.1

RMAs: SOME ILLUSTRATIVE EXAMPLES

RMA	Nature of Paradigm Shift	Core Competency Affected	Dominant Player Affected
Carrier warfare	Created new operational and tactical-level model for naval warfare	Accurate naval gunfire of battleship fleets (rendered obsolete)	Battleship fleets (U.S. and British)
Blitzkrieg	Created new operational and tactical-level model for land warfare	Static defense of prepared positions by infantry and artillery (rendered irrelevant)	French army
ICBM	Created new dimension of warfare (intercontinental strategic warfare)	Long-range, accurate delivery of high-yield nuclear weapons (a new core competency)	
Machine gun	Created new tactical-level model for land warfare	Ability to maneuver massed infantry forces in the open (rendered obsolete)	All armies employing massed infantry forces in the open
Longbow	Created new tactical-level model for land warfare	Man-to-man combat capability of knights on horseback (rendered obsolete)	French armored cavalry

European players in land warfare at the end of World War I, did not develop the blitzkrieg concept of tank warfare, and the British navy, one of the dominant players in sea warfare, did not develop the concept of carrier warfare.[13]

[13]On the other hand, the U.S. Navy, one of the two dominant naval powers in the world at the end of World War I (along with the British navy), did bring about the carrier warfare RMA in the 1920s and 1930s. This is one of the few historical cases (known to the author) of a dominant player developing an RMA. It may tell us something about what it takes for a dominant player (like today's U.S. military) to be successful in transforming its military forces by carrying out its own RMA. For this reason, we will come back to this example in Chapter Six.

The carrier warfare RMA was developed independently by the Japanese navy during the same period. Little is available in English regarding the Japanese development of

- *RMAs frequently bestow an enormous and immediate military advantage on the first nation to exploit them in combat.* A few of many examples are the use of the longbow by the English against the French at Crecy in 1346,[14] the use of the machine gun by the British against the Zulus in 1879 (we discuss this further below), the use of the blitzkrieg by the German army against the Poles in 1939 and the British and French in 1940, and, most recently, the use of stealth aircraft and precision-guided munitions by the United States against the Iraqis in 1991.
- *RMAs are often adopted and fully exploited first by someone other than the nation inventing the new technology.* For example, even though the key inventors of the machine gun were all Americans (William Browning, Richard Gatling, Isaac Lewis, and Hiram Maxim),[15] machine guns were first used in a decisive fashion by European armies against native forces in Africa in the 1870s–1890s.[16] The American army did not begin buying them in quantity and actively incorporate them into its tactical doctrine until many years later,[17] after they were employed by the German army in September 1914 to stop the Allied advance at the Chemin des Dames ridge on the river Aisne.[18] Similarly, the British invented the tank. Although they first employed it in combat during the Battle of the Somme on September 15, 1916 and later at the Battle of Cambrai on November 20, 1917, they

---

carrier aviation. Also, the Japanese navy was not a dominant player at the end of World War I, when its development of carrier aviation began. For these reasons, we do not discuss the Japanese experience in any detail in this report.

[14]The English had developed the technology of the longbow and operational concepts for its use in combat during a long series of civil wars within Britain, but the French had never seen it employed in combat. See Churchill (1958), pp. 332–351.

[15]See Ellis (1975).

[16]One of the first engagements in which machine guns played a decisive role was the Battle of Ulundi, in Natal in 1879, in which a British force equipped with four Gatling guns defeated the Zulu army. (Earlier the same year, a similar size British force without Gatling guns had been virtually wiped out by the same Zulu army at the Battle of Isandhlwana.) See Ellis (1975), pp. 82–84.

[17]It is a little known fact that General George Armstrong Custer's Seventh Cavalry possessed four Gatling guns. Custer left them in garrison when he departed on the campaign that led to Little Big Horn in 1876, since he felt they did not have tactical value (Ellis, 1975, p. 74).

[18]See Ellis (1975), p. 119 and p. 124. The German employment of machine guns from dug-in positions in this battle marked the beginning of World War I trench warfare.

did not understand how to fully exploit its capabilities. This was first shown by the Germans in 1939–1940.[19] Likewise, in 1914 the British conducted the first carrier air raid in history, years before any other navy had operational carriers or carrier-based aircraft. However, they did not develop the RMA of carrier warfare, the American and Japanese navies did, as they demonstrated in the four major carrier battles of 1942.[20]

- *RMAs are not always technology-driven.* For example, American combat tactics during the Revolutionary War (i.e., engaging an opponent from behind cover rather than in formation out in the open) brought about a revolutionary change in land combat without any change in the weapon technologies involved.
- *Technology-driven RMAs are usually brought about by combinations of technologies, rather than individual technologies.* More precisely, technology-driven RMAs are usually brought about by weapons or systems exploiting combinations of technologies. Examples include the blitzkrieg, which was enabled by the combination of three technologies—the tank, the two-way tactical radio, and the dive bomber; and the ICBM, which was enabled by the combination of three technologies—long-range ballistic missiles, lightweight fusion warheads, and highly accurate inertial guidance.
- *Not all technology-driven RMAs involve weapons.* For example, the coming of the railroad to Europe and America in the 1830s–1850s led to a revolution in strategic mobility. This was first demonstrated by the French when they moved 250,000 men at heretofore unheard-of speed to the front in northern Italy to engage the Austrians during the War of 1859. It was later demonstrated (by both sides) on numerous occasions in the 1860s dur-

[19] See Macksey (1975), Corum (1992), and Murray and Watts (1995) for discussion of the invention of the tank and its subsequent exploitation in the blitzkrieg concept.

[20] The battles of the Coral Sea, Midway, the Eastern Solomons, and the Santa Cruz Islands (see Morison, 1963, pp. 140–163, 177–182, and 190–196). See Murray and Watts (1995, pp. 61–84) and Watts and Murray (1996, pp. 383–405) for the steps that led to the carrier warfare RMA, why the Americans "got it," and why the British did not.

ing the American Civil War, and (particularly by the Germans) in 1870 during the Franco-Prussian War.[21]

- *All successful technology-driven RMAs appear to have three components: technology, doctrine, and organization.* Technology, even when developed into a revolutionary weapon or system, is not enough to produce an RMA. It must be combined with doctrine (i.e., an agreed-upon concept for the employment of the new weapon or system)[22] and organization (i.e., a military force structure crafted to exploit the new weapon or system). For example, the blitzkrieg RMA resulted from the combination of the tank, two-way radio, and dive-bomber technologies, an operational concept in which highly mobile armored forces broke through enemy lines and rapidly penetrated to the rear, and a force structure (the panzer division) that concentrated the available tanks into a few specialized divisions.[23] The carrier aviation RMA resulted from the combination of technologies enabling military aircraft to take off and land on carrier decks; the operational concept allowed carrier aircraft to engage an opposing naval force at distances well beyond naval gunfire range and concentrate their attack on the opposing carriers. The force structure (the carrier task force) was built around the aircraft carrier and its planes.[24]
- *There are probably as many "failed" RMAs as successful RMAs.* Some comparatively recent examples include the nuclear-

---

[21]See Brodie (1973, pp. 148–151) and van Creveld (1989, pp. 158–159).

[22]Dupuy (1966) defines doctrine as "Principles, policies, and concepts which are combined into an integrated system for the purpose of governing all components of a military force in combat, and assuring consistent, coordinated employment of these components." Doctrine normally includes concepts of operation, tactics, and, at its fullest, principles of strategy.

[23]In contrast, the French, who had more (and better) tanks in 1940 than did the Germans, spread them out more or less equally throughout all the divisions of the French army (the wrong force structure) and used them as mobile fire support to the infantry (the wrong doctrine). During the 1920s and 1930s, the U.S. Army also viewed tanks primarily as infantry support weapons (the wrong doctrine); this led them to develop tanks with low-velocity guns (the wrong system), which were significantly inferior to the German tanks (with high-velocity guns) they faced in World War II. (See Johnson, 1990 and 1998.)

[24]Dupuy (1984) discusses the critical role that the marriage of new weapons and new doctrine plays in the creation of an RMA.

powered military aircraft, the electromagnetic gun, and the thus-far unfruitful attempts to develop high-energy laser (HEL) weapons for use in military combat.[25] (We will come back to the subject of failed RMAs in Chapter Three.)

RMAs often take a long time to come to fruition. There are many examples of this. The U.S. Navy began experimenting with aircraft in 1910; it took them almost three decades to fully develop the carrier warfare RMA.[26] Similarly, the German army began experimenting with tanks in the early 1920s; it took them almost two decades to create the blitzkrieg.[27] Further back in time, although all of the major technology developments embodied in the machine gun were essentially completed by the 1870s, it did not come to fruition as an RMA in European warfare until September 1914, some 40 years later.[28] Even further back in time, the English developed the technology of the longbow and operational concepts for its use in combat over almost a century of civil wars in Britain, before springing it on the French at Crecy in 1346.[29] So the "revolution" in revolutions in military affairs does not mean the change will occur rapidly—sometimes it will, often it won't—but ultimately it will be profound.[30, 31]

---

[25]See JDR (1986) for discussions of the evolution of HEL application thinking as of the mid-1980s. See APS (1987) for an assessment of the ballistic missile defense applications of HELs. Thus far, all of these attempts to develop militarily useful HEL weapons have been unsuccessful. However, the jury is still out; the latest application focus is on airborne HELs as an antitheater ballistic (ATBM) weapon. (See Aviation Week, 1996.)

[26]We discuss the U.S. Navy's development of the carrier warfare RMA more fully in Chapter Six.

[27]See Guderian (1952), Macksey (1975), and Corum (1992).

[28]See Ellis (1975).

[29]See Churchill (1958), pp. 332–351.

[30]Andrew Marshall (1995) makes this same point in his 1995 writing on RMAs, in which he says: "The term 'revolution' is not meant to insist that the change will be rapid—indeed past revolutions have unfolded over a period of decades—but only that the change will be profound, that the new methods of warfare will be far more powerful than the old."

[31]Some RMAs do happen quickly, however. The best recent example may be the atomic bomb, which was developed and employed over a period of only four years. See Rhodes (1986).

- *The military utility of an RMA is frequently controversial and in doubt up until the moment it is proven in battle.* The British did not begin to realize the combat value of the machine gun until they used it with devastating force against the Zulus at Ulundi in 1879. Many British and French generals continued to seriously doubt the value of machine guns in a European war up until the Germans employed them to stop the Allied advance in September 1914.[32] Not only most French and British generals but many German generals, including some in the German high command, doubted the value of the blitzkrieg up until the moment Guderian broke through at Sedan on May 13–14, 1940, and were vehement in expressing their doubts. Some French, British, and German generals continued to doubt it for days thereafter, even after Guderian reached the English Channel on May 20.[33] Many American admirals seriously doubted the power of carrier aviation up until the battle of Midway in June 1942.[34]

LESSONS FROM THE BUSINESS WORLD REGARDING PARADIGM SHIFTS

Paradigm shifts are not limited to the military arena. They occur in the business world as well, where they have become a much-studied phenomenon.[35] A clear message from the business literature regarding product and process innovation is that product revolutions—the business world's version of paradigm shifts—are rarely brought about by dominant players. According to Utterback (1994):

> Discontinuous innovations that destroy established core competencies . . . almost always come from outside the industry (23 of 29 cases, with 4 from inside and 2 inconclusive).[36]

---

[32]See Ellis (1975).

[33]See Guderian (1952), Macksey (1975), Liddell Hart (1979), and Corum (1992).

[34]See Turnbull and Lord (1949).

[35]See Barker (1992), Utterback (1994), Grove (1996), and Christensen (1997) for four recent examples of this literature. (Grove uses the term "strategic inflection point" and Christensen uses the term "disruptive technological change" to denote the phenomenon, rather than "paradigm shift," but their meanings are the same.)

[36]Utterback (1994, p. 208).

The following list is illustrative:[37]

- Electric typewriters did not come from a major typewriter manufacturer.
- Ballpoint pens did not come from the pen industry.
- Levi's did not come up with designer jeans for women.
- Semiconductors did not come from the vacuum tube industry.
- Radial tires did not come from a major tire maker.
- Personal computers did not come from a major computer manufacturer.
- Wine coolers came from neither the wine nor soda industries.
- Disposable diapers did not come from the diaper services.

The typical impact of these "discontinuous innovations" on dominant players in the business world is stated by Bower and Christensen:

> One of the most consistent patterns in business is the failure of leading companies to stay at the top of their industries when technologies or markets change.[38]

Or in the words of Grove:

> when a strategic inflection point sweeps through the industry, the more successful a participant was in the old industry structure, the more threatened it is by change and the more reluctant it is to adapt to it.[39]

The historical message is clear: in neither military nor business affairs are "revolutions" (i.e., paradigm shifts that destroy core competencies) often brought about by dominant players.

---

[37]Private communication from Samuel Gardiner.

[38]See Bower and Christensen (1995).

[39]See Grove (1996), p. 50.

THE RELATIONSHIP BETWEEN BREAKTHROUGH TECHNOLOGIES AND RMAs

What is the relationship between breakthrough technologies and RMAs? As our previous discussion shows, technology-driven RMAs are brought about by weapons or systems exploiting combinations of technologies, combined with supporting doctrine and organization. Technology alone, without accompanying doctrine and organization, cannot produce an RMA.

Use of the term "breakthrough technologies," therefore, focuses on one of the *inputs* to the breakthrough process; use of the term "RMA" focuses on the *output* from that process. This is the essence of the relationship between breakthrough technologies and RMAs.

We discuss the process that leads from breakthrough technologies to RMAs in Chapter Three.

IS THE CURRENT MILITARY-TECHNICAL REVOLUTION A TRUE RMA?

There is another interesting question regarding RMAs: Is the current military-technical revolution—called by some "the RMA"—a true RMA? Based on our definition, it is too soon to tell. For it to be a true RMA, it must render obsolete or irrelevant one or more core competencies of a dominant player, or create one or more new core competencies in a new dimension of warfare. This has not yet happened.

But it could happen. For example, the use by the U.S. Air Force of air-delivered, precision-guided, antiarmor submunitions in a future regional conflict might conceivably stop the advance of a sizable (e.g., division-size or greater) enemy armored force in its tracks, without requiring intervention by U.S. Army mechanized forces.[40] If this were to occur, and if it could be confidently accomplished in a wide variety of tank-accessible terrain and in the face of enemy air defenses, it would be a true RMA, since it would render irrelevant a core competency (tank/antitank warfare) of a dominant player (the armored forces of the U.S. Army).

---

[40]See Bowie et al. (1993).

As another example, the employment of cyberspace-based techniques by one side in a future conflict might inflict strategic damage on the other side sufficient to significantly alter the course of the conflict.[41] If this were to occur, it would also be a true RMA, since it would create a new core competency (information warfare) in a new dimension of warfare (cyberspace).

Neither of these—nor other examples that have been mentioned in the recent military-technical revolution/RMA literature—has as yet occurred. But because they could, the jury is still out regarding whether the current military-technical revolution will result in one or more true RMAs.

This conclusion is in keeping with Andrew Marshall's initial words regarding what has come to be called "the RMA":

> There is also a tendency to talk about **the** military revolution. This could have the sense that it is already here, already completed. I do not feel that is the case. Probably we are just at the beginning, in which case the full nature of the changes in the character of warfare have not yet fully emerged; therefore, the referent of the phase, "the military revolution," is unclear and indeed should remain to some extent undefined. It would be better to speak about the **emerging** military revolution, or the **potential** military revolution. What we should be talking about is a hypothesis about major change taking place in the period ahead, the next couple of decades. (Emphasis in the original.) (Marshall, 1993.)

Indeed, by prematurely declaring the current military-technical revolution a "revolution in military affairs," the most enthusiastic proponents of "the RMA" may have unnecessarily opened themselves up to criticism.[42]

---

[41]See Molander et al. (1996).

[42]Mann (1998) is but one example of such criticism. Even worse, by terming the ongoing military-technical revolution "the" revolution in military affairs rather than merely "a" revolution in military affairs, as if it were the only RMA that ever occurred, the proponents of "the RMA" show a lack of historical sense.

Chapter Three

THE BREAKTHROUGH PROCESS LEADING TO RMAs

Here again the historical record of technology-driven changes in military operations provides numerous insights into the nature of the breakthrough process leading to RMAs.

RMAs RESULT FROM SERENDIPITOUS CONCEPTUAL BREAKTHROUGHS

One insight is that RMAs almost always involve some sort of conceptual breakthrough that could not be anticipated in advance, and often was not sought for.[1] Based on this insight, Figure 3.1 presents our first and simplest model of the breakthrough process, in which RMAs result from serendipitous conceptual breakthroughs.

In the *preparatory phase*, one or more technology developments and various unmet military challenges[2] set the stage for the subsequent conceptual breakthrough. In the *breakthrough phase*, the key creative event in the RMA process—the critical conceptual breakthrough—occurs. Such conceptual breakthroughs usually cannot be anticipated in advance, and often are not sought for. They often occur accidentally and happen serendipitously.

---

[1]Burke (1978) gives many historical examples of such serendipitous conceptual breakthroughs, in the military as well as in other arenas.

[2]Unpublished 1995 RAND research by Sam Gardiner and Daniel Fox on "Understanding Revolutions in Military Affairs" shows that without one or more unmet military challenges, there is little likelihood of a conceptual breakthrough. The unmet challenges provide a creative impetus essential to the breakthrough process.

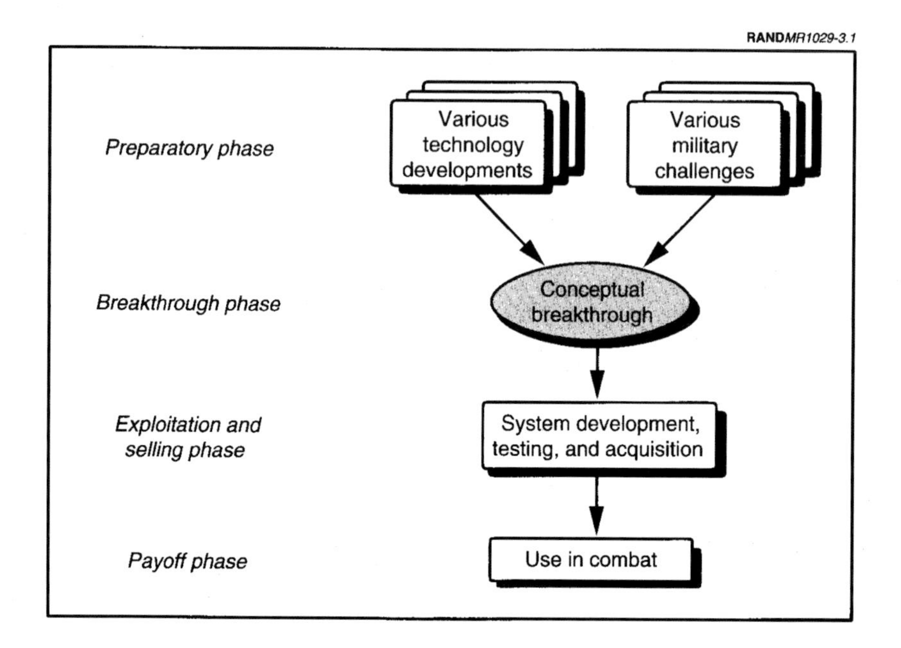

Figure 3.1—One Model of the RMA Process: RMAs Result from Serendipitous Conceptual Breakthroughs

In the *exploitation and selling phase*, the conceptual breakthrough is exploited and sold. It is developed into a military weapon or system, combined with a suitable operational doctrine, and expressed in a force structure adequate to realize the potentialities. It is sold by overcoming the resistance of the many individuals and organizations who can say "no" to the new idea.[3] In the *payoff phase*, the new weapon or system is used in combat and shows its revolutionary potential; the RMA becomes a reality.

RMAs ARE THE RESULT OF MULTIPLE INNOVATIONS

The model in Figure 3.1 is simple, and portrays the accidental nature of the key creative event in the process. However, in suggesting that each successful RMA depends on only one such key innovation, it is

---

[3]We discuss failed RMAs in more detail later, and give examples of such naysayers.

deficient; history shows that each individual RMA is usually the result of a number of innovations in technology, doctrine, and organization. Accordingly, Figure 3.2 presents a second, more complex model of the breakthrough process, in which RMAs are the result of multiple innovations.

The innovative stages in this model are:

- *A new technology* (or several new technologies) that enables devices and systems not previously possible or contemplated.
- *A new device*, based on this new technology, that does something not previously doable.
- *A new system*, based on the new device, that performs a military function either dramatically better or dramatically differently

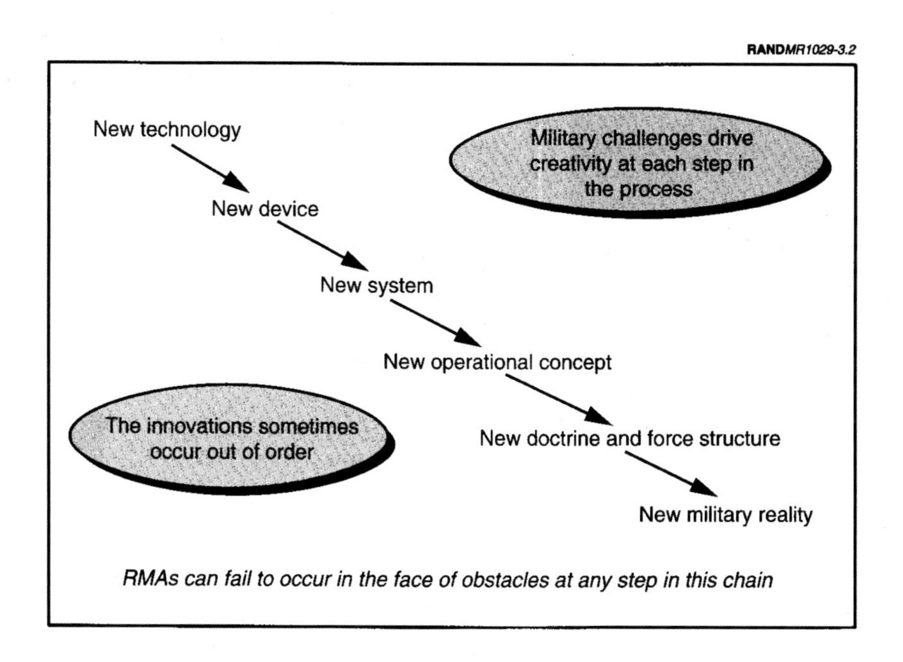

Figure 3.2—Another Model of the RMA Process: RMAs Are the Result of Multiple Innovations

than it had been performed before. (In some cases, it may perform a function that had never been performed before.)

- *A new operational concept* that describes the manner in which the new system is employed in some type of military situation, accomplishing some military task either dramatically better or dramatically differently than it had been accomplished before, or performing a new task that did not exist previously.
- *A new doctrine and force structure*—doctrine that codifies the principles governing the employment of the new system and force structure that provides the military organization necessary to fully realize its potential.

These various stages culminate in *a new military reality,* in which a paradigm shift has occurred in some segment of the military arena.

Figure 3.2 also shows that

- Unmet military challenges are an essential element driving creativity at each step in the process. Without one or more challenges, technologies are unlikely to be combined into devices and devices into systems, and new operational concepts, doctrine, and force structures are unlikely to be developed.
- The various innovations sometimes occur out of order: e.g., an operational concept is "invented" before a technology, device, and/or system exists adequate to realize its postulated potential.[4]
- RMAs can fail to occur in the face of obstacles at any step in this chain. The necessary technology may exist, but the contemplated devices prove impractical. It may not be possible to turn the new devices into viable systems. No operational concept may exist to employ an otherwise viable system concept. The force structure to exploit the operational concept may not exist because the operational concept is unacceptable to the pre-

---

[4]For example, the operational concept of strategic bombardment was developed during the 1920s and 1930s. (See Mitchell, 1925, for a discussion by one of its early proponents.) However, the aircraft and weapon technologies were not robust enough to support an RMA. It was only with the development of nuclear weapons, intercontinental-range bombers, and ICBMs in the 1940s–1960s that an RMA resulted.

vailing military culture, or because the new force structure requires too large a change in existing military organizations.

We will return to the subject of failed RMAs later in this chapter.

MUCH OF THE RMA PROCESS CAN BE OBSERVED AND ANTICIPATED

The model in Figure 3.1 emphasizes the serendipitous nature of (at least some) key creative events in the RMA process. The model in Figure 3.2 emphasizes the multiple innovations that make up the process. Both of these models deal with the internals of the RMA process. The model in Figure 3.3, on the other hand, deals with an aspect of the externals of the process: the signals that can be seen by outside observers.

Serendipitous invention that is the essential creative element at the heart of the RMA process and leads from the new technology to the new device, operational concept, and system concept is difficult to

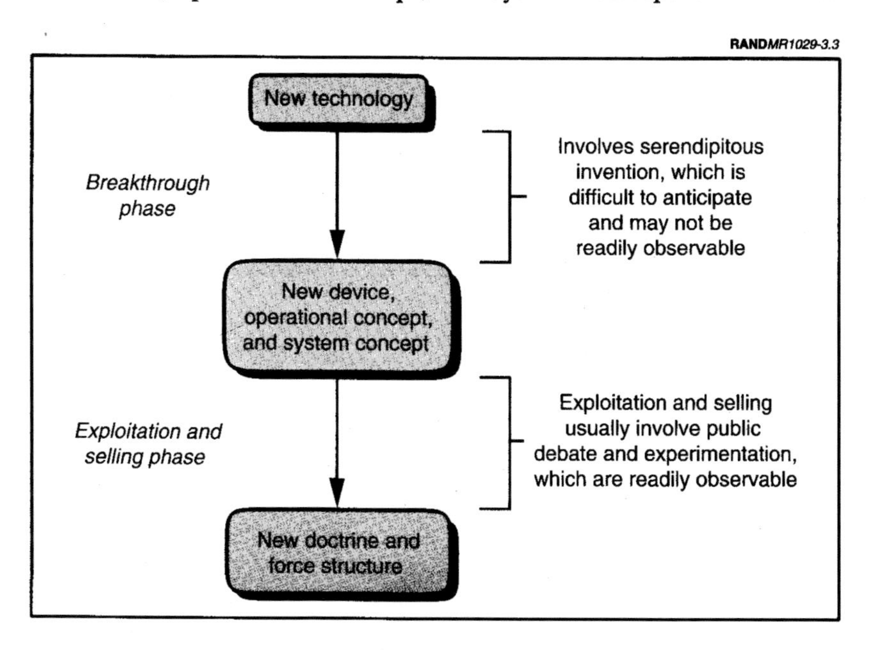

Figure 3.3—Much of the RMA Process Can Be Observed and Anticipated

anticipate and may not be readily observable, particularly at the moment it occurs. However, during the exploitation and selling phase that leads from the new device, operational concept, and system concept to a new doctrine and force structure—and which is absolutely essential if a new doctrine and force structure are to be achieved—there is usually public debate and experimentation that are readily observable. Thus, the latter stages of the RMA process are almost always accompanied by readily observable signals, which anticipate the emergence of new RMAs.[5] We discuss these observables in more detail later.

MUCH CAN BE LEARNED FROM FAILED OR INCOMPLETE RMAs

Additional insights concerning the RMA process can be gleaned from a more detailed look at the history of some failed or incomplete RMAs and why they failed to achieve their anticipated potential. As indicated earlier, RMAs can fail to occur in the face of obstacles at any step in the chain portrayed in Figure 3.2. We next consider historical examples associated with each of these possibilities.

Necessary Technology Exists But Contemplated Devices Prove Impractical

In the 1950s, much thought was given to a nuclear-powered aircraft, which would have virtually unlimited range and endurance, and would therefore (in its proponents' view) revolutionize aerial warfare.

This dream never came to fruition. Even though the necessary nuclear reactor and energy conversion technologies existed, the contemplated device (a nuclear-powered aircraft engine) proved much too heavy to be practical. Because of the weight of the engine, such an aircraft would literally have never gotten off the ground.

---

[5]In the 20th century, the author knows of only two RMAs that were not preceded by readily observable, public signals: the atomic bomb and the stealth aircraft (and the jury is still out as to whether the stealth aircraft is a true RMA).

New Devices Cannot Be Turned Into Viable Systems

More recently, in the 1970s and early 1980s, thought was given to an electromagnetic (EM) gun that would shoot projectiles at much higher muzzle velocities than conventional guns, and would therefore (in its proponents' view) be a superior antitank, antiaircraft, and antimissile weapon.

In this case, EM guns—or rather EM accelerators that accelerated small projectiles to velocities of several kilometers per second—were in fact developed and tested, and performed (as a device) more or less as their proponents had claimed. However, even though these EM guns worked in principle, in practice they were cumbersome, with internal barrel components that wore out rapidly and had to be replaced often (sometimes after every shot). For these reasons (and probably others as well) it has, thus far at least, not been possible to turn EM guns into viable military systems.

No Operational Concept Exists to Employ an Otherwise Viable System Concept

Without an operational concept, the best weapon system in the world will never revolutionize anything. The machine gun—or rather the lack of a position for the machine gun—in most European-based armies during the last quarter of the 19th century provides a good example of this. By 1885, the development of a workable machine gun was relatively complete, and several firms were actively marketing such guns. But most European armies—with the exception of the British (they missed the full significance of this RMA for a different reason)—did not have the slightest concept of how to employ these guns effectively in combat. In the battles of Wissembourg and Spichern in 1870 during the Franco-Prussian War, the French tried using machine guns mounted on artillery carriages and sited with the field artillery as indirect fire weapons.[6] They were outranged by the Prussian artillery pieces and blown to bits before they had a chance to fire. None of the other leading European-based armies (except the British, about which more later) came up with a better idea during the period before 1900. The idea that the use of

[6]See Ellis (1975), pp. 63–64.

machine guns as direct-fire infantry-support weapons could decimate infantry forces attempting to cross open ground did not occur to them.[7]

Wrong or Incomplete Operational Concept Is Used

Sometimes one or more seemingly small missing elements in an operational concept can cause the failure of an RMA, or can cause one player to miss realizing the full potential of an RMA that another player achieves. Carrier warfare provides an example. On Christmas Day 1914 the British conducted the first carrier air raid in history—the attack on the Cuxhaven Zeppelin base near Wilhelmshaven by seven British seaplanes from three improvised carriers in the Heligoland Bight.[8] At the close of World War I, the Royal Navy had over three years of wartime carrier operations and possessed nearly a dozen carriers of one sort or another, at a time when no other naval power had even one.[9]

In spite of this head start, the British completely missed realizing the full potential of the carrier warfare RMA; at the beginning of World War II, the first-line British carriers were incapable of generating the combat striking power of American and Japanese carriers, as so convincingly demonstrated in the carrier battles of 1942. Why did the British miss this RMA? For the seemingly smallest of reasons. Their concept of operations did not include the "deck park," the practice of stowing a major fraction of a carrier's complement of aircraft on the flight deck, and refueling and rearming there as well. The British stowed, refueled, and rearmed all of their aircraft below, on the

---

[7]It did occur to the Russians and later to the Japanese, who both employed machine guns effectively during the Russo-Japanese War of 1904–1905. German observers of this war took the idea back home, where the German army adopted it and began adding machine guns to its forces, thus setting the stage for their employment to stop the Allied advance on the river Aisne in September 1914. See Ellis (1975), pp. 65–68.

[8]Three troop carriers, *Engadine, Riveria,* and *Empress,* modified to carry small numbers of seaplanes, conducted this raid. For flight operations, the seaplanes were lowered into the water using cranes. See Murray and Watts (1995), Watts and Murray (1996), and Friedman (1988).

[9]These vessels ranged from early seaplane carriers, such as *Empress* and *Riveria,* to *Ark Royal,* the first ship designed and built as an aircraft carrier, and *Argus,* the first flat-deck carrier. See Murray and Watts (1995), Watts and Murray (1996), and Friedman (1988).

hanger deck. Thus, in 1939 a first-line British carrier carried only 24–30 aircraft, whereas American and Japanese carriers carried 80–100 aircraft.[10] It turned out that the key determinants of the offensive striking power of a carrier force were the number of strike aircraft that could be launched in a single attack and how quickly successive attacks could be mounted. Because the American and Japanese carriers carried many more aircraft, they could launch much larger attacks; because they refueled and rearmed their aircraft on the flight deck, they could turnaround returning aircraft much faster, thereby launching more and faster successive attacks. These features made all the difference in the world.[11]

The development of tank doctrine by the American army during the 1920s and 1930s is another example of the consequences of a wrong or incomplete operational concept. The U.S. Army viewed tanks primarily as infantry support weapons, an incomplete operational concept that ignores the possibility and importance of tank-versus-tank engagements. This led them to develop tanks with low-velocity guns (the wrong system), which were significantly inferior to the German tanks (with high-velocity guns designed to go against other tanks) they faced in World War II.[12]

No Doctrine and Force Structure to Exploit the Operational Concept Because the Concept Is Unacceptable to Prevailing Military Culture

Sometimes both a viable system and an effective operational concept exist, but because the operational concept is unacceptable to the prevailing military culture, the doctrine and force structure necessary to exploit the new weapon are not developed. This was the case regarding the use of the machine gun in the British army during the period leading up to World War I.

---

[10]The British carriers were somewhat smaller than the American and Japanese carriers, which also limited the number of aircraft they could carry. But the biggest limiting factor was their lack of deck parks.

[11]See Murray and Watts (1995), pp. 61–84, and Watts and Murray (1996), pp. 383–405.

[12]See Johnson (1990 and 1998).

In contrast to the continental European armies, by the 1880s the British knew how to employ machine guns in combat to achieve devastating effect: as direct-fire infantry weapons. The British learned this in Africa, fighting the native tribes. Machine guns were used against the Zulus at Ulundi in 1879, in the assault on Tel-el-Kebir in Egypt in 1882, against the Dervish at Abu Klea in the Sudan in 1884, and again against the Dervish at the Battle of Omdurman in 1898.[13]

But these were the British colonial forces, not the mainstream British army; and these were native tribes, not other "civilized" European armies. Simply put, the prevailing British military culture could not conceive of "officers and gentlemen" employing such an uncivilized weapon against other officers and gentlemen. In the words of Ellis:

> So the machine gun became associated with colonial expeditions and the slaughter of natives, and was thus by definition regarded as being totally inappropriate to the conditions of regular European warfare.[14]

Thus, in the years before World War I, the British army did not develop the doctrine and force structure necessary to exploit the machine gun.[15]

No Force Structure to Exploit Operational Concept Because the New Force Structure Requires Too Large a Change in Existing Military Organizations

Sometimes the force structure necessary to exploit a viable system and a recognized operational concept requires too large a change in existing military organizations, and is therefore not developed. This was the case regarding the development—or rather the arrested

[13]See Ellis (1975), pp. 82–86. As but one example of the devastating effect of machine guns in these engagements, in the Battle of Omdurman 11,000 Dervish were killed, brought down primarily by six Maxim guns; on the British side, only 28 British soldiers and 20 other (colonial) soldiers were killed.

[14]Ellis (1975), p. 57.

[15]See Ellis (1975), pp. 48–60.

development—of tank warfare in the British army during the 1920s and 1930s.[16]

As mentioned earlier, the British invented the tank and were the first to employ it in combat, during World War I. Following that war, a number of British individuals (most prominently J.F.C. Fuller and B. H. Liddell Hart) wrote and spoke passionately regarding the tank's potential to revolutionize land warfare, laid out operational concepts to that end, and advocated a new force structure for the British army centered on all-tank units. Further, the British army carried out an innovative series of experiments in the late 1920s and early 1930s involving the use of armor in mobile, mechanized warfare. The most notable of these experiments were the 1926 maneuvers on the Salisbury plain, in which an armored force carried out a 25-mile penetration that wrecked the defending forces' position.[17]

In spite of the apparent lessons of these maneuvers and the passionate arguments of armored-warfare advocates such as Fuller and Liddell Hart,[18] the leaders of the British army rejected this new operational concept and the force structure that went along with it. The new force structure proposed by Fuller, Liddell Hart, and their followers required too large a change in the then-existing organizational structure of the British army; it upset too many apple carts and provoked too much opposition from defenders of traditional regiments. In the words of Murray and Watts:

> The path of British innovation in armor . . . remained outside the army's mainstream, and the educational process that the experi-

---

[16]See Murray and Watts (1995), pp. 25–30, for a detailed discussion of the aborted British attempts during this period to develop a doctrine and force structure fully exploiting the tank.

[17]German observers were present at these 1926 maneuvers and carried the (apparent) message regarding the tank's operational potential back home, where it was picked up by forward-looking thinkers in the German army. (See Murray and Watts, 1995, pp. 18–30.)

[18]In fact, the increasingly vehement and strident arguments of Fuller, Liddell Hart, and their followers may well have been part of the problem; they tended to polarize the debate and antagonize the mainstream British military leaders. (See Murray and Watts, 1995, pp. 25–30.)

ments with armor might have developed into became a "we versus them" contest between old and new.[19]

> The split of the [British] army into two separate camps (with the radical innovators, by far, the smaller) insured that [the radical innovators'] ideas played little if any role in the preparation of the British army for war in the late 1930s.[20]

Thus, in the years before World War II, the British army did not develop the doctrine (mobile, mechanized warfare) and force structure (armored divisions) necessary to exploit the tank, and thereby missed out on the blitzkrieg RMA.

Force Structure and Operational Concept Not Congruent with Grand Strategy

The French also failed to adopt a doctrine of offensive tank warfare during the period between the World Wars, but for a different reason: they were focused on a grand strategy for land warfare that was primarily defensive. The enormous casualty lists of World War I trench warfare had convinced the leaders of the French army that in the future all offensive operations, except those that were limited and tightly controlled, would no longer be worth the price. They could not conceive of the tank overcoming the power of the defense that had been demonstrated in 1914–1918. Accordingly, in the 1920s and 1930s they adopted a land warfare doctrine that was almost entirely defensive. Moreover, and more important, they could not conceive of any other (successful) way to fight. Doughty describes

> the fundamental unwillingness and inability of senior French military leaders to accept the possibility that others might wage future war in a fashion very different from theirs.[21]

Offensive tank operations had no place in the French strategy, so the French also missed out on the blitzkrieg RMA.

---

[19]See Murray and Watts (1995), p. 28.

[20]See Murray and Watts (1995), p. 29.

[21]Doughty (1985 and 1990), as paraphrased by Murray and Watts (1995, p. 25).

These historical examples of failed or incomplete RMAs reinforce some of the characteristics of RMAs discussed in Chapter Two.

- Successful technology-driven RMAs require technology, doctrine, and organization. Missing or incomplete elements in any one of these areas can cause a military force to "miss out" on an RMA.
- RMAs are rarely brought about by dominant players, because such players are often not motivated to make the necessary doctrinal or organizational changes.[22]
- RMAs are often adopted and fully exploited by someone other than the nation originally inventing the new technology, because that nation's military failed (for whatever reason) to make the necessary doctrinal or organizational changes.

They also highlight some additional lessons regarding the RMA process:[23]

- Military institutions must be willing to develop a *vision* of how war may change in the future, or they are incapable of developing RMAs.
- Acceptance of new ideas by (at least some) senior military leaders and by (at least part of) the military bureaucracy is essential to the successful development of RMAs by existing military institutions.[24]

---

[22]As mentioned earlier, the U.S. Navy, one of the dominant naval players at the end of World War I, did develop the carrier warfare RMA during the 1920s and 1930s. We discuss this in Chapter Six.

[23]Murray and Watts (1995, pp. 84–93) and Watts and Murray (1996, pp. 405–415) highlight these additional lessons.

[24]As stated by Murray and Watts (1995, p. 87), "it seems unlikely that any small handful of visionaries, however dedicated and vocal, have much chance of forcing military institutions to adopt fundamentally new ways of fighting without the acquiescence or grudging cooperation implied by emerging bureaucratic recognition and acceptance." See also Watts and Murray (1996, p. 409). As an example of the successful harnessing of this bureaucratic process, Rosen (1991) discusses the key role played by a few senior naval leaders in facilitating the development of the carrier warfare RMA by the U.S. Navy during the 1920s and 1930s.

- The potential for civilian or outside leadership to impose a new vision of future war (i.e., the vision of an RMA) on a reluctant military service whose heart remains committed to existing ways of fighting is, at best, limited.
- Institutional processes for exploring, testing, and refining conceptions of future war—i.e., for conducting experiments and assessing their results—are essential to the development of RMAs. (The German army and the U.S. Navy's aviation community had such processes during the 1918–1939 period; the British and French armies did not.)

We now turn to a discussion of how one may be prepared for future RMAs—carried out by others.

PART II. BEING PREPARED FOR FUTURE RMAs (CARRIED OUT BY OTHERS)

Chapter Four

BEING AWARE OF THE NEXT RMA: THE OBSERVABLES OF THE EMERGENCE OF NEW RMAs

The first step in being prepared for future RMAs carried out by others is *being aware* that an RMA may be occurring. As mentioned in conjunction with Figure 3.3, much of the RMA process can be observed and anticipated. This is particularly true during the exploitation and selling phase (see Figure 3.3) that leads from a new device, operational concept, and system concept to a new doctrine and force structure—and which is absolutely essential if the new doctrine and force structure required to truly realize an RMA are ever to be achieved. We discuss these observables in this chapter and describe the essential elements of an activity to monitor and assess such observables on a continuing basis.

THE RMA PROCESS PRODUCES OBSERVABLES IN A NUMBER OF VENUES

During the exploitation and selling phase of the RMA process, observables are produced in:

- *Various press organs,* including the trade press (defense, aerospace, etc.), the military art and science press, the science and technology press, the international security and foreign affairs press, and the general business press, as well as leading newspapers and magazines.

  Trade press organs covering the defense and aerospace arena include *Asia-Pacific Defense Forum, Asian Defence Journal, Asia-Pacific Defence Reporter, Aviation Week & Space Technology, Canadian Defence Quarterly, Defense & Aerospace Electronics,*

Defense News, Defense Week, International Defense Review, Jane's Defence Weekly, Military Technology, and *Signal.*[1]

Military art and science press organs include *Air Force Magazine, Air Force Times, Airpower Journal, Armed Forces Journal International, Armor, Army, Army Times, Field Artillery, Infantry, Marine Corps Gazette, Military Review, Naval War College Review, Navy International, Navy Times,* and *Proceedings of the United States Naval Institute.* Science and technology press organs include *Nature, Science, Scientific American,* and *Technology Review.*

The international security and foreign affairs press includes *Defense & Foreign Affairs, European Security, Foreign Affairs, International Affairs, NATO Review, NATO's Sixteen Nations, Strategic Review,* and *The Journal of Strategic Studies.* General business press organs include *Barron's, Business Week, Forbes, Fortune, The Economist, The Financial Times,* and *The Wall Street Journal.* Leading newspapers and magazines include *Der Spiegel, Die Welt, Die Zeit, Frankfurter Allgemeine, International Herald Tribune, Neue Zürcher Zeitung, The New York Times, The Times* (of London), and *The Washington Post.*

Because the emergence of any "true" RMA is almost certain to provoke considerable public debate (because of the established military institutions whose position it threatens) and have significant business and economic impact (because of the changes it implies in the defense industry), it is highly unlikely that such an RMA would not be covered in any of these press organs. It is bound to show up somewhere.

- *The worldwide arms market,* both legitimate and clandestine. Arms dealers and their salesmen tout their products worldwide. Their activities could be useful indications of an emerging technology-driven RMA.
- *Inferior military establishments* trying to leapfrog the dominant players. Often, such military establishments are among the first to try out a new and distinctly different military system/

---

[1]This list and the ones that follow are meant to be illustrative; they are most certainly not exhaustive.

operational concept that offers the hope of rendering obsolete a core competency of a dominant military player. Monitoring the activities of such lesser military establishments may lead to early indications of an emerging RMA.

- *Dominant military players* trying to discredit new ideas that threaten their core competencies. The focus here should not be on the new military systems/operational concepts that the dominant military players are pursuing, but rather the new systems/concepts that the dominant players are arguing against—and the more vehemently they are arguing against something new and potentially revolutionary, the more likely it is that this might indeed be an emerging RMA.
- *Military research, development, test, and evaluation (RDT&E) activities,* particularly those involving new technologies, systems, and/or operational concepts.[2]

Any process designed to be on the lookout for emerging technology-driven RMAs should watch all of these venues, both within and outside the United States.[3]

OPEN AND CLOSED VENUES REQUIRE DIFFERENT COLLECTION APPROACHES

Some of these venues are open, some are closed. The activities in open venues are usually readily observable by (almost) anyone; the activities in closed venues are normally shut off from view by outsiders. As indicated in Figure 4.1, the legitimate arms market is generally an open venue; its activities are normally readily observable. (Indeed, frequently they are actually advertised.) The unclassified

---

[2]An illustrative example of the potential usefulness of this venue is that of the Germans who observed the 1926 British maneuvers on the Salisbury plain and carried the message regarding the tank's operational potential for mobile warfare back home to the intellectual leaders of the (then embryonic) German army.

[3]Unpublished 1997 RAND work by Jeffrey Isaacson, Christopher Layne, and John Arquilla presents a number of predictors on whether a state is likely to achieve military innovation. Such predictors can help focus this process.

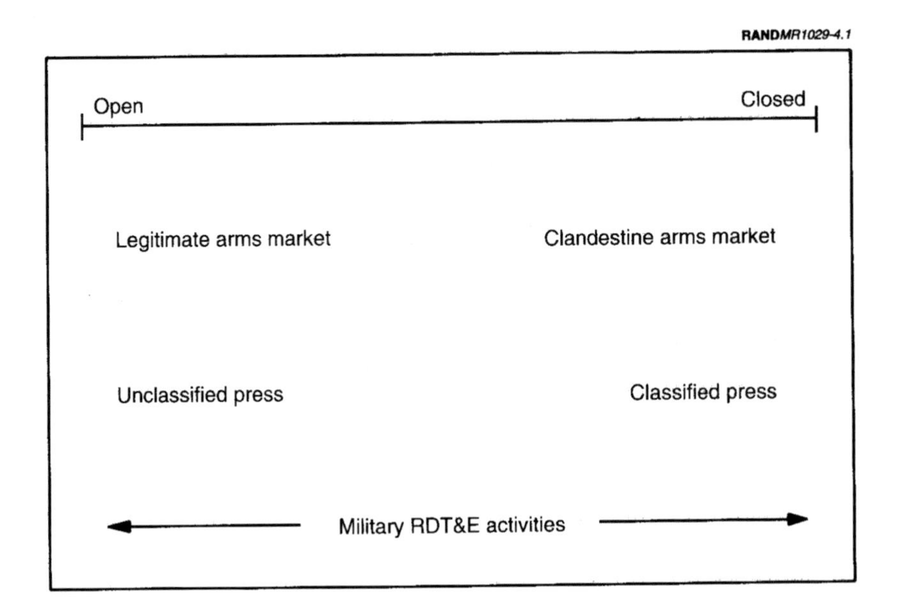

Figure 4.1—Some of These Venues Are Open, Some Are Closed

press is an open venue. The clandestine arms market, on the other hand, is a closed venue; participants in this market go to great lengths to hide their activities from general view. The classified press is also a closed venue, accessible only to those with the proper security clearances.

Open and closed venues require different collection approaches. For open venues, the standard open-source collection techniques—surveying newspapers, periodicals, and books, monitoring television and radio broadcasts, attending conferences, etc.—are applicable, albeit tailored to targets of specific relevance to the RMA process. For closed venues, standard human intelligence (humint) and communications intelligence (comint) techniques are applicable, most likely targeted based on open-source cueing information.

COLLECTION IS NOT ENOUGH; ASSESSMENT IS ALSO REQUIRED

As the discussions in Chapters Two and Three have shown, not all potential RMAs come to pass; many are aborted and fall by the wayside for a variety of reasons. Accordingly, the collection of observables emanating from the emergence of new RMAs is not enough; these observables must also be carefully assessed, to separate out the serious RMA candidates from the wild-eyed dreams. Figure 4.2 illustrates what is needed in this assessment process.

As shown in the figure, a multistep collection and assessment process is required, with the following components:

- *An initial, wide-area-search collection process,* to detect any and all RMA visions and dreams, no matter where they arise and no matter how far-out they may appear. The emphasis here should be on inclusion rather than exclusion. The output of this continually ongoing collection activity is a living list of RMA visions and dreams.[4]
- *An initial screening process,* based on some sort of plausibility criteria, to weed out the "antigravity" ideas[5] (or their equivalent) from this list but keep in all those with some prospects of success. At this stage in the assessment, it is much safer to keep questionable ideas in than to throw good ideas out; the plausibility criteria used should be selected accordingly. The output of this step is a list of potential RMA candidates.
- *A monitoring collection process,* focused on each of the potential RMA candidates and continuing over an extended period. The specifics of this collection will vary, depending on the nature of each potential RMA, and may focus on specific challenges, hurdles, or tests that a given candidate RMA must pass.
- *A careful assessment process,* which could include challenges, hurdles, and tests that a candidate RMA must pass.[6] The output

---

[4]A living list may be changing all the time, or at least every year or so.

[5]"Antigravity" ideas are concepts that are clearly not feasible, based on fundamental physical or engineering considerations.

[6]The specifics will vary, based on the nature of each RMA candidate.

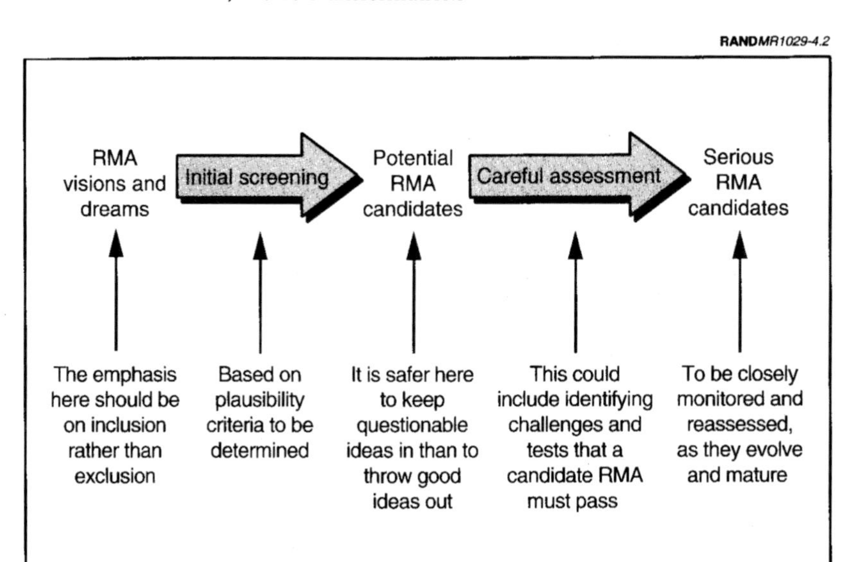

Figure 4.2—Since All Potential RMAs Do Not Pan Out, Collection Is Not Enough: Careful, Balanced Assessment Is Also Required

of this ongoing step is a list of serious RMA candidates, to be closely monitored and reassessed as they evolve and mature.

This process requires patience and staying power. Since future RMAs cannot be scheduled and may take years to come to fruition, one must establish a collection and assessment process that can endure over a long period .

IN ASSESSING POTENTIAL BREAKTHROUGHS, DO NOT DEPEND ON EXPERTS ALONE

In conducting both the initial screening and careful assessment steps in this process, one cannot rely on the views of experts alone; they cannot always foresee the future. Table 4.1, taken from Cerf and Navasky (1984), gives examples of cases where experts "got it wrong."[7] Table 4.2, from the same source, shows that military experts similarly cannot always foresee the military future.

[7]Cerf and Navasky (1984) have over 300 pages of such examples.

Table 4.1

Experts Can't Always Foresee the Future

"The phonograph . . . is not of any commercial value."
Thomas Alva Edison, inventor of the phonograph, c. 1880

"Heavier-than-air flying machines are impossible."
Lord Kelvin, British mathematician, physicist, and president of the British Royal Society, c. 1895

"Man will not fly for fifty years."
Wilbur Wright to his brother Orville, 1901

"I cannot imagine any condition which could cause a ship to founder. . . . Modern shipbuilding has gone beyond that."
Captain Edward J. Smith, White Star Line (future commander of the Titanic), 1906

"With the possible exception of having more pleasing lines to the eye while in flight, the monoplane possesses no advantages over the biplane."
Glen H. Curtiss (Founder of Curtiss Aircraft), December 31, 1911

"Who the hell wants to hear actors talk?"
Harry M. Warner (President of Warner Brothers Pictures), c. 1927

"A severe depression like that of 1920–1921 is outside the range of probability."
The Harvard Economic Society, 16 November 1929

"I think there is a world market for about five computers."
Thomas J. Watson, chairman of IBM, 1943

"We don't like their sound. Groups of guitars are on the way out."
Decca Recording Co. executive, turning down the Beatles in 1962

"With over fifty foreign cars already on sale here, the Japanese auto industry isn't likely to carve out a big slice of the U.S. market for itself."
Business Week, 2 August 1968

"There is no reason for any individual to have a computer in their home."
Ken Olson, president, Digital Equipment Corporation, 1977

SOURCE: Christopher Cerf and Victor Navasky, *The Experts Speak*, Pantheon Books, New York, 1984.

Table 4.2

Nor Can Military Experts Always Foresee the Military Future

"Make no mistake, this weapon will change absolutely nothing."

French Director General of Infantry, dismissing (before members of the French parliament) the importance of the machine gun in warfare, 1910

"[The machine gun is] a grossly overrated weapon."

British Field Marshal Douglas Haig, at the outbreak of World War I, c. 1914

"It is highly unlikely that an airplane, or fleet of them, could ever sink a fleet of Navy vessels under battle conditions."

Franklin D. Roosevelt (former Assistant Secretary of the Navy), 1922

"As for tanks, which are supposed by some to bring us a shortening of wars, their incapacity is striking."

Marshal Henri Philippe Pétain (former French Minister of War and former Commander-in-Chief of the French Armies), 1939

"Their [the German] tanks will be destroyed in the open country behind our lines if they can penetrate that far, which is doubtful."

General A. L. Georges (Major-Général des Armées), 1939

"There are no urgent measures to take for the reinforcement of the Sedan sector."

General Charles Huntziger (Commander of the French Second Army), May 13, 1940

"No matter what happens, the U.S. Navy is not going to be caught napping."

Frank Knox, Secretary of the Navy, 4 December 1941, just before the Japanese attack on Pearl Harbor

"This is the biggest fool thing we have ever done.... The bomb will never go off, and I speak as an expert in explosives."

Admiral William D. Leahy, advising President Harry S. Truman on the impracticality of the U.S. atomic bomb project, 1945

"They couldn't hit an elephant at this dist..."

General John B. Sedgwick (Union Army), last words, uttered during the Battle of Spotsylvania, 1864

SOURCE: Christopher Cerf and Victor Navasky, *The Experts Speak*, Pantheon Books, New York, 1984.

The message of Tables 4.1 and 4.2 is (at least) twofold, insofar as the assessment of potential RMAs is concerned:

- One cannot always depend on the views of experts regarding the prospects for revolutionary change.
- The views of experts can be particularly unreliable when they have a stake in the old way of doing things.

Both the initial screening and careful assessment steps in the process outlined in Figure 4.2 must be structured with these admonitions in mind.

THE ESSENTIAL ELEMENTS OF A WORLDWIDE RMA BREAKTHROUGH WATCH AND ASSESSMENT ACTIVITY

We have identified the essential elements of what we term a "worldwide RMA breakthrough watch and assessment activity":[8]

- An *information collection activity* that conducts two types of collection:
  - — Worldwide search, primarily open-source, to uncover new RMA visions.
  - — Continued monitoring, using open-source techniques and (if necessary) standard closed-source techniques (e.g., Humint and Comint), focused on RMA candidates that have survived the initial screening process.
- An *RMA assessment activity* that conducts two types of assessment:
  - — Initial screening, to weed out the equivalent of anti-gravity ideas but keep in all those items with some prospects of success.

---

[8]Eugene Gritton (RAND) and David Signori (then at the Defense Advanced Research Projects Agency [DARPA] and now at RAND) suggested this term to the author.

— Continued and more careful assessment, over time, to follow potential/serious RMA candidates as they evolve and mature and see if they surmount various challenges and hurdles.

As mentioned earlier, since one never knows when a future RMA may arise or come to fruition, these collection and assessment activities must be established in such a way that they can endure over a long period. They can be carried out in two separate but closely coupled organizations—information collection in some sort of intelligence organization and RMA assessment in some sort of an advanced military research and development organization—or they can be carried out in an organization having combined capabilities. In either case:

- The organization carrying out the information collection activities must be able to access open sources of information (newspapers, periodicals, books, television and radio broadcasts, conferences, etc.) on an effective, worldwide basis, and must also be able to call upon closed-source humint and comint collection techniques where needed.
- The organization carrying out the RMA assessment activities must have access to creative thinkers with expertise in science, technology, military systems, and military operations—creative thinkers who can combine "out of the box" thinking with an appreciation for practical realities.[9]

This worldwide RMA breakthrough watch and assessment activity, if properly implemented in an enduring fashion, should ensure U.S.

---

[9]DARPA is an obvious candidate to be the organization entrusted with RMA assessment activities. In fact, it performs somewhat similar functions today, albeit focused in a number of areas DARPA management has identified as potentially revolutionary insofar as U.S. military operations are concerned, rather than encompassing the entirety of RMA "dreams and visions" throughout the world, whether or not they would apply to U.S. military forces and the current U.S. military strategy. DARPA certainly has access to creative, "out of the box" thinkers with expertise in science, technology, military systems, and military operations.

DARPA also has a history of close and fruitful interactions with intelligence agencies skilled in accessing open sources of information and in conducting closed-source collection operations.

awareness of future RMAs being carried out by others.[10] But being aware of emerging RMAs is not enough; one must also be responsive. We turn to that challenge in Chapter Five.

---

[10]In addition to alerting the United States to RMA threats being developed by others (in time for the United States to prepare countermeasures), the RMA breakthrough watch can also generate RMA opportunities for the United States to develop itself (as the Germans did after observing the 1926 British tank maneuvers on the Salisbury plain).

Chapter Five

BEING RESPONSIVE TO THE NEXT RMA: THE CHARACTERISTICS OF A FUTURE-ORIENTED MILITARY ORGANIZATION

Being adequately prepared to cope with an emerging RMA being developed by others is a twofold challenge:

- *Being aware* of a potential emerging RMA
- *Being responsive* to the implications of that RMA.

Failure to meet either one of these challenges can lead a nation to military disaster. The Zulus were unaware of the machine gun RMA before the Battle of Ulundi, which led them to disaster in that battle. The British and French armies were aware of the blitzkrieg RMA well before the events of May 1940, but failed to respond; this led them to disaster in the Battle of Flanders and the subsequent Battle of France.

Chapter Four dealt with the first of these challenges, describing the essential elements of a worldwide RMA breakthrough watch and assessment activity designed to ensure awareness of future RMAs. This chapter deals with the second.

OVERCOMING THE OBSTACLES TO RESPONSIVENESS

Established military organizations more often than not fail to respond adequately to emerging RMAs threatening their core competencies, even ones of which they are aware, primarily because of inherent obstacles to the changes necessary to cope with an RMA. This can be thought of in terms of obstacles in the path of each of the steps in the RMA process; we took this viewpoint in Chapter Three.

It can also be thought of, more generally, in terms of generic psychological obstacles to the organizational learning and change necessary to cope with paradigm shifts, no matter what their shape or form;[1] we will take this viewpoint here.

Psychological obstacles to change are as common for business organizations confronting paradigm shifts as for military organizations confronting RMAs. Andrew Grove (1996, p. 124), the co-founder and former CEO of Intel, lays out the typical steps in an organization's response to a paradigm shift threatening one of its core competencies:

- Denial
- Escape or diversion
- Acceptance and pertinent action.

How does a military establishment cope with organizational denial when confronted with a potential RMA? How does a military establishment cope with organizational escape or diversion in the face of a potential RMA? How does a military establishment achieve acceptance and pertinent action in response to a potential RMA? We address these questions next.

Overcoming Denial

Psychologists tell us that the first stage in an individual's response to the death of a loved one is almost always denial: psychological denial that the person is gone. The same is true for military organizations threatened with the forthcoming "death" of a cherished core competency (core competencies are a military organization's "loved ones").

Recent history is full of examples of military organizations that were aware of an emerging RMA but failed to respond, most often because of denial. In the period before World War I, the leaders of the infantry and cavalry forces of most European armies were aware of the

[1]We use "psychological" in the sense of organizational psychology—i.e., the behavior of organizations.

machine gun and what it had accomplished against native armies in Africa, but they denied the possibility that it would be used in combat between civilized armies in Europe, as well as the possibility that it could overcome the morale of properly trained infantry or the charge of properly motivated cavalry. In the period before World War II, the leaders of the British and French armies were aware of the claims of the proponents of what became the blitzkrieg RMA, but they denied its efficacy. The list goes on and on.

How does one overcome such organizational denial? According to Grove (1996, pp. 1–3):

> Only the paranoid survive. Sooner or later, something fundamental in your business world will change.
>
> When it comes to business, I believe in the value of paranoia. Business success contains the seeds of its own destruction. The more successful you are, the more people want a chunk of your business and then another chunk and then another until there is nothing left. I believe the prime responsibility of a manager is to guard constantly against other people's attacks [on his organization's core competencies] and to inculcate this guardian attitude in the people under his or her management.

This attitude of "productive paranoia"—our term, not Grove's—is just as applicable to successful military organizations as to successful business organizations. A basic sense of productive paranoia regarding the future is a useful first step in overcoming organizational denial, particularly for a dominant player such as the U.S. military.

But that's not all that is required. According to Murray and Watts (1995, p. 85), based on their analysis of instances of military innovation (and noninnovation) during the period between World War I and II:

> The evidence . . . attests, first of all, to the importance of developing visions of the future. Military institutions not only need to make up-front intellectual investments to develop a vision of future war, but they must continue agonizing over that vision, struggling to discern how the next war may differ from the last.

Murray and Watts (1995, p. 85) and Watts and Murray (1996, p. 407) recount the importance of General Hans von Seeckt's "post–World War I vision of mobile warfare by a highly professional, well-trained, well-led army" to the subsequent development of the blitzkrieg concept by the German army.[2]

The combined message from the business and military arenas is that denial (of change) can be overcome by maintaining a basic sense of productive paranoia regarding the future, and by developing and continually refining a vision of how the future (i.e., future wars) may differ from the past. Two useful techniques for the development of such visions of future wars are wargaming, as employed by Gardiner and Fox,[3] and the concept of asymmetric strategies, described in Bennett et al. (1994a, 1994b, 1998, 1999).[4,5]

Overcoming Escape or Diversion

In the business world, escape or diversion is often the next step in an organization's response to an oncoming paradigm shift. Turning again to Grove (1996, pp. 124–125):

> Escape, or diversion, refers to the personal actions of the senior manager. When companies are facing major changes in their core business, they seem to plunge into what seem to be totally unre-

---

[2]General von Seeckt was head of the German general staff and chief of the German army during the period 1919–1926. Corum (1992) makes this same point.

[3]In their 1995 unpublished RAND work on "Understanding Revolutions in Military Affairs," Sam Gardiner and Daniel Fox conducted an extensive series of wargames exploring six future wars in Southwest Asia. The challenge-response cycle in this series of future Gulf wars—first one side gains the advantage, then the other—generated a continually evolving vision of how future wars may differ from previous wars.

[4]The term "asymmetric strategies" denotes a certain class of military strategies (or operational concepts) employed by an opponent of a dominant military player. These strategies are asymmetric in the sense that they do not mimic the dominant player's approach to warfare. Rather, they deliberately choose a different way of conducting combat—a way chosen to negate the dominant player's many advantages.

[5]Other "futuring techniques" recently employed in the business or military arenas which may be useful here include "scenario-based planning" (Schwartz, 1991), "microworlds" (Senge, 1990), "future search" (Weisbord and Janoff, 1995), "assumption-based planning" (Dewar, 1993), "discovery-driven planning" (McGrath and MacMillan, 1995, and Christensen, 1997), and "The Day After . . ." methodology (Molander et al., 1996).

> lated [activities]. In my view, a lot of these activities are motivated by the need of senior management to occupy themselves respectably with something that clearly and legitimately requires their attention day in and day out, something that they can justify spending their time on and make progress in instead of figuring out how to cope with an impending strategically destructive force.

Military organizations facing paradigm shifts are often subject to the same phenomena.

How does one overcome such organizational escape or diversion? Developing and continually refining a vision of how future wars may differ from past wars, mentioned above as a means of overcoming denial, will certainly help here also. Broad and intensive debate regarding the future of the organization is also of value. In the words of Grove (1996, p. 99):

> How do we know whether a change signals a strategic inflection point [Grove's term for a paradigm shift]? The only way is through the process of clarification that comes from broad and intensive debate.

The message is clear: escape or diversion can be overcome by developing and continually refining a vision of how future wars may differ from past wars, and by fostering an organizational climate encouraging broad and intensive debate regarding the future of the organization.

Achieving Acceptance and Pertinent Action

Overcoming denial, escape, and diversion in the face of an emerging RMA is not the end of the story. The organization must then unite behind an effective response to the challenge (what Grove calls "acceptance and pertinent action"). According to Grove (1996, p. 121):

> Resolution comes through experimentation. Only stepping out of the old ruts will bring new insights.

Or, in the words of Murray and Watts (1995, p. 88), based on their analysis of military innovation between World War I and II:

> Institutional processes for exploring, testing, and refining conceptions of future war . . . are literally the *sine qua non* of successful military innovation in peacetime.

In both the business and military arena, there must be mechanisms available within the organization for experimentation with new ideas, even if they threaten the organization's current core competencies.

More is required to achieve acceptance and pertinent action. Rosen's investigation of the politics of peacetime innovation in 20th century military organizations shows a need for (at least some) senior officers with traditional credentials who sponsor the new ways of doing things (within at least part of the organization).[6] In Rosen's words (p. 76):

> Innovations occurred when senior military officers were convinced that structural changes in the security environment had created the need. These senior officers, who had established themselves by satisfying the traditional criteria for performance, had the necessary power to champion innovations.

General von Seeckt played this role during the initial stages of development of the blitzkrieg RMA;[7] Admiral William S. Sims (the president of the Naval War College during 1917–1922)[8] and Rear Admiral William A. Moffett (the director of the U.S. Navy's Bureau of Aeronautics from 1921 to 1933) played the same role during the early developmental stages of the carrier aviation RMA.[9]

In addition, new promotion pathways (within at least part of the organization) for junior officers practicing a new way of war are also necessary. In the words of Rosen (1991, p. 251):

---

[6]See Rosen (1991), pp. 76–105.

[7]See Corum (1992).

[8]In April 1917, Admiral Sims was called away from his Naval War College position to become Commander, U.S. Naval Forces in Europe, in preparation for the U.S. entry into World War I. Sims returned to the Naval War College in December 1918, where he remained until his retirement in 1922. (See Murray and Watts, 1995, pp. 69–70.)

[9]See Turnbull and Lord (1949), Melhorn (1974), and Murray and Watts (1995, pp. 19–22 and 69–74).

> Peacetime innovation has been possible when senior military officers with traditional credentials, reacting . . . to a [perceived] structural change in the security environment, have acted to create a new promotion pathway for junior officers practicing a new way of war.

In summary, three things appear to be necessary to achieve acceptance and pertinent action in a military organization confronted with an emerging RMA: mechanisms within the organization for experimentation with new ideas, senior officers willing to sponsor new ways of doing things, and new promotion pathways for junior officers practicing a new way of war.

THE CHARACTERISTICS OF A FUTURE-ORIENTED MILITARY ORGANIZATION

The characteristics of a future-oriented military organization likely to respond adequately to an emerging RMA include:

- "Productive paranoia" regarding the future.
- A continually refined vision of how war may change.
- An organizational climate encouraging vigorous debate regarding the future of the organization.
- Mechanisms available within the organization for experimentation with new ideas, even ones that threaten the organization's current core competencies.[10]
- Senior officers with traditional credentials willing to sponsor new ways of doing things.

---

[10]As a recent study by the CNO [Chief of Naval Operations] Executive Panel shows, this usually requires separation of the revolutionary innovative activities from the mainstream activity of the military organization—i.e., in separate, nonbureaucratic organizations. R. Robinson Harris (CAPT, USN), Executive Director, CNO Executive Panel, "Naval Warfare Innovation," briefing to RAND, August 5, 1998; and Thomas Tesch, staff member, CNO Executive Panel, "Naval Warfare Innovation Task Force," briefing to the CNO Executive Panel, June 16, 1998.

Christensen (1997) makes the same point regarding "disruptive" product innovation in business organizations; to be successful it must be carried out in specially created organizations separate from the mainstream.

- New promotion pathways for junior officers practicing a new way of war.

Possessing these characteristics is no guarantee of future success. However, a military establishment lacking one or more of these characteristics is less likely to respond adequately to an emerging RMA.

There is a final challenge: These characteristics must come from within the military establishment in question; they cannot be imposed from the outside. As Murray and Watts (1995, p. 87) concluded based on their case studies of military innovation in the 1920s and 1930s:

> The dynamics evident in the case studies suggest that the potential for civilian or outside leadership to *impose* a new vision of future war on a reluctant military service whose heart remains committed to existing ways of fighting is, at best, limited.

Thus, this future-oriented military organization must be within the military establishment in question (i.e., the U.S. Army, Navy, Air Force, or Marine Corps) rather than outside.

PART III. BRINGING ABOUT FUTURE RMAs (OF YOUR OWN)

Chapter Six

WHAT DOES IT TAKE TO BRING ABOUT A SUCCESSFUL RMA?

What does it take for a military organization to bring about an RMA of its own, rather than merely responding to an RMA being developed by someone else? History suggests that all of the following items are probably necessary:

- You must have a fertile set of enabling technologies.
- You must have unmet military challenges.
- You must focus on a definite "thing" or a short list of "things."
- You must ultimately challenge someone's core competency.
- You must have a receptive organizational climate
  - — that fosters a continually refined vision of how war may change and
  - — that encourages vigorous debate regarding the future of the organization.
- You must have support from the top
  - — senior officers with traditional credentials willing to sponsor new ways of doing things
  - — new promotion pathways for junior officers practicing a new way of war.
- You must have mechanisms for experimentation
  - — to discover, learn, test, and demonstrate.

- You must have some way of responding positively to the results of successful experiments
  - — in terms of doctrinal changes, acquisition programs, and force structure modifications.

In what follows we expand on each of these items.

YOU MUST HAVE A FERTILE SET OF ENABLING TECHNOLOGIES

For a technology-driven RMA, you must have a fertile set of enabling technologies.[1] It helps greatly if these technologies are new and/or emerging rather than old and mature: new and emerging technologies are much more likely to be fertile breeding grounds for revolutionary developments than old, mature technologies, whose "revolutions" are usually well in the past. Indeed, military history is full of RMAs in the years immediately following major advances in technology. Table 6.1 gives a few examples.

YOU MUST HAVE UNMET MILITARY CHALLENGES

Unmet military challenges are essential elements driving creativity at each step in the RMA process. Without one or more challenges, technologies are unlikely to be combined into devices and devices into systems; and new operational concepts, doctrine, and force structures are unlikely to be developed.

Inferior military establishments, particularly those that lost the last war, are usually well supplied with unmet military challenges—*obvious* unmet challenges—that can serve as a driving force for substantial change. This may not be the case for superior military establishments, particularly those that won the last war. Such military organizations frequently feel on top of the world, with no need for change, certainly not radical change. Special attention is therefore required to motivate the RMA process.

---

[1]As mentioned in Chapter Two, not all RMAs have been techology-driven. But most of them have been, and the focus of recent RMA-related discussions and of this report is on technology-driven RMAs.

Table 6.1

The Technologies Behind Some RMAs

RMA	Period of Development	Enabling Technology	Years of Emergence[a]
ICBM	1955–1965	Fusion weapons	1950–1955
		Multistage rockets	1945–1955
		Inertial guidance	1950–1955
Atomic bomb	1941–1945	Nuclear fission	1938
Carrier warfare	1921–1939	Aviation	1900–1915
		Radio communications	1900–1915
Blitzkrieg	1921–1939	Tanks	1915–1918
		Radio communications	1900–1915
		Dive bombing	1921–1926

[a]The years of emergence are (approximately) when each technology first appeared. Each continued to develop and mature for many years.

YOU MUST FOCUS ON A DEFINITE "THING" OR A SHORT LIST OF "THINGS"

Fertile enabling technologies by themselves are not enough. They must come together in a definite "thing": a device or system exploiting the enabling technologies together with a concept for operational employment. Table 6.2 illustrates this, showing the device/system and employment concept "things" involved in several RMAs.

The more fertile a set of enabling technologies, the more possibilities it offers for combinations of devices, systems, and employment concepts, and the more of a challenge it is to focus down on the right combination of device, system, and employment concept to bring about an RMA. This focusing process can take considerable time; until it occurs there is no RMA.

The U.S. Navy's experience in developing the carrier warfare RMA is illustrative. Beginning in the early 1910s, the Navy experimented with a number of air vehicles: seaplanes, flying boats, planes with wheels, and three types of lighter-than-air vehicles (rigid airships, blimps, and kite balloons); a number of different basing concepts for

Table 6.2

The "Things" Involved in Some RMAs

RMA	Device/System	Employment Concept
ICBM	Long-range ballistic missile with fusion warhead and inertial guidance	Bombardment of strategic, fixed targets
Carrier warfare	Wheeled planes operating from fast, flat-deck ships	Airborne attack of naval surface targets
Blitzkrieg	Tanks, two-way radios, and dive bombers	Mobile maneuver warfare
Machine gun	Rapid fire, anti-personnel gun	Direct-fire weapon against massed infantry formations

these air vehicles: airfields on land, seaplane tenders, ships with catapults, and ships with flat decks for landing and takeoff; and a variety of missions: scouting, spotting (the fall of naval gunfire), air defense (of the fleet), attack of land targets (e.g., naval bases), and attack of naval targets (e.g., ships at sea). All of these air vehicles were made possible by the evolving aviation technology, as were all of the basing concepts. And all of the different missions seemed of value to some part of the U.S. Navy.

It took the Navy over 20 years to experiment with the different combinations of air vehicle, basing concept, and mission application and finally concentrate on wheeled planes, based on fast, flat-deck ships that could keep up with the battle fleet wherever it went (i.e., fleet aircraft carriers), to be used primarily to attack naval targets and secondarily for air defense of the fleet—that is, the carrier warfare RMA, which finally emerged in the 1930s and was proven in combat in 1942.[2]

[2]See Turnbull and Lord (1949), Melhorn (1974), Murray and Watts (1995), and Watts and Murray (1996) for detailed discussions of the U.S. Navy's development of carrier aviation.

YOU MUST ULTIMATELY CHALLENGE SOMEONE'S CORE COMPETENCY

RMAs are all about core competencies: creating new ones and upsetting old ones. To create an RMA, you have to challenge an existing core competency of a dominant military player. This is natural (albeit possibly difficult) for a nondominant player to do; it is not so natural for a dominant player to do, because it may have to challenge one of its own core competencies—it may have to render obsolete something that makes it a powerful, superior military organization.

As we said earlier, history is full of examples of inferior military powers developing RMAs that overcome a superior opponent. In many of these cases, the inferior power deliberately set out to upset a core competency of its superior opponent. The German army under General von Seeckt deliberately set out in the 1920s to overcome the core competency of the French army (demonstrated during World War I) for static defense of prepared positions by infantry and artillery; they succeeded and created the blitzkrieg RMA.[3] Somewhat earlier, the German navy developed and exploited the U-boat (during World War I) as a counter to the dominant (surface) naval power of the British navy. Still earlier, in the early 1800s, Napoleon combined the *levée en masse* (the mobilization of mass armies), the *grande batterie* (the physical massing of artillery), the "attack column," and several other tactical and operational innovations into an overall system of war to overcome the Prussian army, then dominant in European warfare.[4] Much earlier, in the 1300s, the English deliberately set out to overcome the numerically superior French army's core competency for man-to-man combat by knights on horseback by exploiting (in a tactical system/operational concept) the longbow technology they had developed during a series of civil wars in Great Britain.

Again, there are few historical examples of a superior military power developing an RMA that upsets one of its core competencies. The development of carrier warfare by the U.S. Navy is the only clear example known to this author. In this case, the developers of U.S. naval aviation rendered obsolete the core competency of the U.S.

---

[3]See Corum (1992).

[4]See Dupuy (1984), pp. 154–168.

battleship force for accurate, overwhelming naval gunfire, a core competency that made the U.S. Navy one of the two dominant naval warfare players in the world (along with the British navy) at the end of World War I.

However, the people who developed U.S. naval aviation did not set out to challenge the battleship's core competency; they initially set out to support the battleship force—to provide scouting support, gunfire spotting support, and air defense support to the battleship force, so that it could continue to dominate and defeat an opposing battleship force, any opposing battleship force, in surface combat.[5] It was only over time that they came to realize that carrier aviation could replace the battleship force as the principal combat force of the U.S. Navy.[6]

You do not have to start out challenging someone's core competency, although that is the usual historical pattern. But ultimately you have to mount such a challenge, or you will not have an RMA.

---

[5]In the early 1920s, Brig. Gen. William Mitchell of the Army Air Service was the apostle of those who believed that air power would make the battleship obsolete. (See Mitchell, 1921 and 1925). The leaders of the then-fledgling naval aviation did not subscribe to Mitchell's view. They brought about the creation of the Bureau of Aeronautics in 1921 explicitly to develop naval aviation in support of the battleship force, not to eliminate the battleship. (See Turnbull and Lord, 1949, and Melhorn, 1974.) This primary focus of naval aviation as a support to the battleship force continued throughout the 1920s and well into the 1930s. As late as 1938, the Navy version of War Plan Orange, the plan for military operations against Japan in the Pacific, envisaged carriers accompanying the battleship force to provide scouting and air defense support as it fought its way across the Pacific to regain the Philippines. (See Melhorn, 1974.)

[6]Some naval aviators came to this realization earlier than others. Among the earliest were the staff of Aircraft Squadrons, Battle Fleet (COMAIRONS), who in 1928 began thinking about the possibility of carrier task forces operating independently of the battleship force conducting offensive operations (including attacking opposing battle fleets). The COMAIRONS staff planned the first such independent carrier task force operation, conducted during the 1929 fleet exercise: The carrier USS *Saratoga* (CV-3) accompanied by the cruiser *Omaha* broke off from the main Red force off the coast of southern Mexico, steamed 660 miles southeast, and then northeast around the opposing Blue force over a 24-hour period, past the Galapagos Islands, along the north coast of South America, and into the Gulf of Panama. Then 150 miles off of the Panama Canal (Red's objective in the fleet exercise), it launched a 66-plane air strike that (theoretically) destroyed the locks of the Canal and heavily damaged Army airfields in the Canal Zone. (See Wilson, 1950.)

YOU MUST HAVE A RECEPTIVE ORGANIZATIONAL CLIMATE

We said in Chapter Five that you need a receptive organizational climate to respond to someone else's RMA. You also need one to develop your own. You need an organizational climate that encourages vigorous debate regarding the ways in which war may change and the impact of those changes on the military organization in question. You need a climate that encourages change, wants the organization's future to be different from its past, and that wants and welcomes change. You need this receptive organizational climate even more if the RMA in question, in this case internally rather than externally generated (the case in Chapter Five), threatens a core competency of the organization. It takes a brave organization to make a part of itself obsolete. Historically, this has been rare in the military world.

It has also been rare in the business world—rare but not unknown. There have been a few companies that for periods of time have had deliberate policies of making their leading products (i.e., their core competencies) obsolete, of making them obsolete before someone else did, and had organizational climates supporting these policies.[7] These examples show that it can be done, that such organizational climates can exist.

In a nutshell, if a military organization wants to bring about a successful RMA, it must have a receptive organizational climate. If the military organization is a dominant player and wants to bring about an RMA that upsets one of its core competencies, it must have a *very* receptive organizational climate.

YOU MUST HAVE SUPPORT FROM THE TOP

We also said in Chapter Five that you need support from the top to respond to someone else's RMA. You need support from the top to develop your own, too. You need (at least) two types of support: senior officers with traditional credentials willing to sponsor new ways of doing things, and new promotion pathways for junior officers practicing a new way of war. Both are essential if an RMA is to occur.

---

[7]Intel is a recent example of such a company. (See Grove, 1996.)

The U.S. Navy had both of these during the period between the two World Wars. Several Navy admirals provided essential support at crucial periods during the development of carrier aviation, most notably including Admiral William S. Sims and Rear Admiral William A. Moffett. Promotion of naval aviators to the rank of commander and captain was a problem in the early years, but from the mid-1930s on, all captains commanding carriers and naval air stations had to be qualified naval aviators; this provided a promotion pathway to higher ranks.[8]

The U.S. Army, on the other hand, did not have high-level support during the interwar period for changes in the way it waged war, particularly changes involving new ways of employing tanks or aircraft. Not only did the generals commanding the traditional branches of the Army (infantry, artillery, and cavalry) oppose the development of innovative ways of using tanks and aircraft, they also put promotion roadblocks in the way of any officers persisting in careers in the fledgling Armored Corps or Army Air Corps.[9]

In sum, the innovators in the U.S. Navy had support from the top during the interwar period; they produced the carrier warfare RMA. The innovators in the U.S. Army lacked such support; the Army entered World War II with both armor and aviation doctrine and technology markedly inferior to that of the Germans.[10]

YOU MUST HAVE MECHANISMS FOR EXPERIMENTATION

To bring about an RMA, a military organization must have mechanisms available within the organization for experimentation with new ideas, to discover, learn, test, and demonstrate:

---

[8]See Turnbull and Lord (1949).

[9]See Johnson (1990 and 1998) for a detailed discussion of how the U.S. Army "got it wrong" regarding the development of innovative armor and aviation doctrines and technologies during the 1920s and 1930s. As but one of many examples of promotion roadblocks put in the way of junior Army officers wanting to pursue new ways of war, both Dwight Eisenhower and George Patton were advised to transfer out of the Armored Corps if they ever wanted to make major. (Private communication from David E. Johnson, 1998.)

[10]See Johnson (1990 and 1998).

- To *discover* what you can do with new technologies and combinations of new technologies, what new devices and systems become possible, what existing military tasks can be done differently, what new military tasks become achievable, what works and what does not.
- To *learn* what combinations of device and system parameters work best, what operational concepts the new devices and systems support well and which ones they do not support well, which operational concepts appear more promising and which less promising, what works better and what doesn't work as well, what makes sense and what does not.
- To *test* promising device, system, and operational concept combinations in a wide variety of real-world circumstances, thereby focusing on the combination of device(s), system(s), and employment concept(s) most likely to bring about an RMA.
- To *demonstrate*, finally, that the chosen set of device(s), system(s), and operational concept(s) offers the potential for a revolutionary improvement in military capabilities in real-world conflict situations.

These experimental mechanisms must allow one to take risks and fail (from time to time), particularly in the earlier *discover* and *learn* stages but also in the *test* stage. It is through taking risks, failing from time to time, and thereby learning what does not work that the necessary focusing discussed earlier is accomplished.

These experimental mechanisms must be available even for new ideas that threaten the organization's current core competencies. This can be difficult. It usually requires separation of the experimental activities involving revolutionary innovation from the mainstream activity of the military organization—i.e., in some separate, nonbureaucratic organizations.[11]

---

[11]The U.S. Navy's CNO Executive Panel makes this point strongly in its recent investigation of innovation in naval warfare. See footnote 10 in Chapter Five. "Nonbureaucratic" is the term it uses to describe the required organizational attribute.

In the business arena, Christensen (1997) makes a similar point regarding the process of discovery required to find markets for new products that threaten to disrupt the mainstream product lines of a company; to be successful these market discovery activ-

The U.S. Navy's development of carrier aviation followed the pattern outlined here. The early naval aviation experiments were carried out separate from the mainstream Navy—in a number of temporary organizations during the 1910s and by the Bureau of Aeronautics from 1921 on—and featured a great deal of *discovery* (of what worked and what did not) and *learning* (what works better and what less well, what makes sense and what does not) in the early stages. It was not until fleet exercises beginning in 1929 that innovative carrier aviation experiments (of the *test* and *demonstrate* variety) were carried out in conjunction with mainstream Navy activities.[12]

YOU MUST HAVE SOME WAY OF RESPONDING POSITIVELY TO THE RESULTS OF SUCCESSFUL EXPERIMENTS

Even if everything works—you have a fertile set of technologies, you focus on a good idea that challenges someone's core competency, you have a receptive organizational climate and support from the top, you have mechanisms for experimentation (even for an idea that threatens one of your organization's core competencies), and the experiments successfully demonstrate the technical feasibility, operational utility, and potentially revolutionary impact of your RMA candidate—you still need more before you have an RMA. You need ways of responding positively to successful experiments in terms of doctrinal changes, acquisition programs, and force structure modifications. These are the final three hurdles.

Doctrinal Changes

Doctrinal changes to accommodate and fully exploit the new device or system are an essential element of any successful RMA.[13] Future-oriented military organizations have mechanisms for periodically rethinking, refining (in small ways), and revising (in big ways) their operational doctrine. RMAs usually require major doctrinal changes.

---

ities must be carried out in specially created organizations separated from the mainstream.

[12]See Turnbull and Lord (1949) and Melhorn (1974) for details of the 20-plus years of Navy experiments that led to the carrier warfare RMA.

[13]As indicated earlier, military doctrine normally includes agreed-upon concepts of operation, tactics, and principles of strategy.

One might assume that the rethinking leading to RMA-related doctrinal changes normally comes after successful experiments have proven the potential of the prospective RMA; this has not always been so. In the case of carrier aviation, Admiral Sims began a series of strategic and tactical exercises at the Naval War College in 1919 to explore the use of aircraft in naval operations. These War College gaming exercises continued throughout the 1920s and 1930s, in parallel with the Bureau of Aeronautics' development of naval aviation technology and experiments with systems, basing concepts, and operational concepts. They led to an evolving series of doctrines for the employment of naval aviation, culminating in the carrier warfare doctrine of 1941–1942. In this case, the doctrinal rethinking was carried out in parallel with the RMA experiments.[14]

In at least one case, the German development of the blitzkrieg, the doctrinal rethinking took place largely before the RMA experiments. Under the leadership of General Hans von Seeckt between 1919 and 1926, the German army developed a doctrine of mobile, maneuver warfare that emphasized combined arms and independent action by commanding officers at all levels; it was designed to regain primacy for the offense (in contrast to the defense dominance of World War I). This doctrinal development was largely complete by 1926 when von Seeckt stepped down as head of the German army. Over the next ten years, the Germans proceeded to develop the devices/systems (the tank, two-way tactical radio, and dive bomber) and force structure (the panzer division) to bring this doctrine to fulfillment in the blitzkrieg.[15]

The U.S. Navy's development of carrier warfare and the German army's development of the blitzkrieg are cases where the doctrinal changes necessary to realize an RMA were accomplished in a straightforward fashion. As we discussed in Chapter Three, many potential RMAs have failed because of doctrinal hurdles. In the 1920s and 1930s, for example, the American and British advocates of new land warfare paradigms exploiting the tank were unable to bring

---

[14]Turnbull and Lord (1949) and Melhorn (1974) discuss the central role that the Naval War College gaming exercises initiated by Admiral Sims played in the Navy's development of carrier warfare doctrine.

[15]See Corum (1992) for a detailed discussion of von Seeckt's role in developing the conceptual foundations for the blitzkrieg.

about the necessary doctrinal changes because of opposition from leaders of the traditional branches (infantry, artillery, and cavalry) of the U.S. and British armies.[16]

Doctrinal changes probably most often come after successful RMA experiments, although sometimes they come in parallel with the experiments, and on at least one occasion they came before the RMA experiments—but they always must come, or there will not be an RMA.

Responsive Acquisition Programs

A military acquisition system able to respond positively to radical innovations is another necessary element of any successful RMA.[17] The challenge here is how the military service's acquisition system deals with *risk*, specifically risk associated with uncertainties regarding the military need and utility of the candidate RMA. Such uncertainties often persist up until the moment the RMA is proven in battle.[18]

Typically, military acquisition systems are set up to deal with new equipment embodying evolutionary improvements in military capability, operating within the well-defined context of existing doctrine (i.e., operational concepts, tactics, and strategy). Such improvements usually involve limited and well-understood risk, of both a technical and military-utility nature, and meet a well-recognized military need. Operating in this environment, present-day military acquisition systems are normally designed to be risk adverse, taking as few chances as possible and usually requiring that all uncertainties (regarding technical issues, military need, and military utility) are resolved before full-scale production begins.[19]

---

[16]Johnson (1990 and 1998) discusses the U.S. Army's lack of doctrinal innovation during the 1920s and 1930s.

[17]By "military acquisition system" we mean the totality of rules, regulations, processes, and procedures governing the acquisition of new equipment for the military service in question.

[18]Candidate RMAs often involve technical risks, but they are usually resolved earlier in the development process than are the military-utility risks.

[19]The DoD 5000 Series' policies and procedures is an example of such a risk-adverse acquisition system. See DoD (1996 and 1997).

An RMA presents a difficult situation for such an acquisition system, even after the RMA's potential has been demonstrated in a successful experiment. While the principal technical issues may have been resolved, the military need and utility of the new RMA system may still be in doubt, or at least highly controversial, because it involves a new concept of operation, new tactics, and perhaps even a new strategy—all untested in battle. The usual acquisition system, with its numerous risk-reduction/elimination milestones, offers numerous opportunities for disbelievers in or opponents of the RMA to impede the acquisition process by raising questions regarding uncertainties and risk.[20]

What is required for a military acquisition system to respond effectively and efficiently to an RMA is that a branch of the acquisition system be set up to handle novel and radical innovations. This RMA branch of the acquisition system should tolerate substantial military-utility risks to a much later stage in the acquisition process, in some cases all the way to initial operational capability (IOC). The stages and decision points in this new acquisition branch should be constructed in keeping with the likely uncertainties at each stage in the RMA process, with an emphasis on fostering novel/radical innovations rather than avoiding mistakes and saving money.[21]

Force Structure Modifications

Finally, after everything else is accomplished, the force structure of the military organization in question must be modified to accommodate new units equipped with the new devices and systems, operating according to the new doctrine. In the case of the carrier warfare RMA, the new and essential force structure element was the carrier task force, organized around the aircraft carrier with supporting cruisers and destroyers. The concept of a carrier task force was first

---

[20]If the new system is at all novel (and RMAs are), there will always be uncertainties and disbelievers. If the candidate RMA challenges any core competency of the military organization in question, there will always be opponents.

[21]Birkler et al. (1999) propose that just such a branch be added to the DoD acquisition process, to handle novel and radical innovations—whether or not they qualify as RMAs.

tried out by the U.S. Navy during the 1929 fleet exercise.[22] Following its success in that exercise, carrier task forces were played in most subsequent fleet exercises during the 1930s. By 1941 the carrier task force was a recognized element of the Navy's combat organization.[23]

In the case of the blitzkrieg RMA, the new and essential force structure element was the panzer division, made up of a combined-arms team of tanks, motorized infantry, artillery, reconnaissance troops, engineers, and support and supply units. The panzer division was developed in the 1930s by Heinz Guderian, on the doctrinal and equipment foundations established by von Seeckt and his followers in the 1920s.[24]

To a large extent, the carrier task forces in the U.S. Navy in December 1941 (prior to Pearl Harbor) and the panzer divisions in the German army in September 1939 (prior to the invasion of Poland) and May 1940 (prior to the invasion of The Netherlands, Belgium, and France) were *add-ons* to the previously existing force structures, not replacements for major elements of those structures. The bulk of the German army in 1939–1940 and the bulk of the U.S. Navy in 1941 were organized in "pre-RMA" units.[25]

This is typical of the history of past RMAs. Most often, the new RMA elements are treated as add-ons to the existing force structure until the candidate RMA has been proven in battle.[26] Major replacements

---

[22]See Wilson (1950).

[23]As late as 1940–1941, many of the U.S. Navy's leaders continued to view the carrier task forces' primary wartime mission as support to the battleship force, not independent operations. Following Pearl Harbor, however, the carrier task forces had to wage independent operations against the Japanese navy; essentially all of the U.S. battleships in the Pacific had been sunk or heavily damaged.

[24]See Guderian (1952), Macksey (1975), and Corum (1992). This was one of Guderian's two key contributions to the blitzkrieg RMA; the other was his operational leadership in Poland (September 1939) and Flanders (May 1940).

[25]For example, the German force invading Poland on September 1, 1939, included six panzer divisions, four light panzer divisions, and 48 old-fashioned infantry divisions; the German force invading The Netherlands, Belgium, and France on May 10, 1940, included 10 panzer divisions and 126 infantry divisions. (See Churchill, 1948, pp. 442–443; and Churchill, 1949, pp. 29–31.)

[26]The carrier task force add-ons to the U.S. Navy's force structure in the 1930s were financially possible because of the expenditure increases initiated by the Roosevelt Administration beginning in 1934 as economic stimulants. If the budget policies of

of old force structure elements by RMA elements do not occur until after that has happened.

* * *

With a fertile set of enabling technologies, unmet military challenges, focus on a definite "thing" or a short list of "things," a challenge to someone's core competency, a receptive organizational climate, support from the top, mechanisms for experimentation, and ways to respond positively to successful experiments in terms of doctrinal changes, acquisition programs, and force structure modifications—and at least one "brilliant idea"—a military organization has a reasonable chance of bringing about a successful RMA. Without any one of these elements, the chances are much less, even if there is a brilliant idea, and history suggests the RMA process is likely to fail.

---

the Harding, Coolidge, and Hoover administrations had been continued in the 1930s, these force-structure add-ons would not have been possible. (See Turnbull and Lord, 1949, pp. 284–295.)

Chapter Seven

DOD'S CURRENT FORCE TRANSFORMATION ACTIVITIES: DOES ANYTHING APPEAR TO BE MISSING? WHAT CAN BE DONE TO FILL IN THE MISSING ELEMENTS?

Since publication of the 1997 *Quadrennial Defense Review* (QDR) (Cohen, 1997), the DoD has been involved in a concerted effort to "transform" the U.S. military, motivated by a fourfold set of objectives:[1]

- To achieve the operational goals outlined in *Joint Vision 2010* (JV2010) (dominant maneuver, precision engagement, full-dimensional protection, focused logistics) (Shalikashvili, 1996; and Joint Staff, 1997)
- To bring about the cost savings necessary to pay for force modernization[2]
- To achieve a new, affordable force structure that can be maintained in the future (in the words of the QDR, "more capability for less money")[3]
- To take advantage of the [so-called] revolution in military affairs currently ongoing—"the RMA."

---

[1]Davis et al. (1998) point out that force transformation is also necessary to meet future military challenges, "already visible and certain to worsen."

[2]Much of the capital equipment (e.g., planes, tanks, ships) of the U.S. military is aging and gradually wearing out. Sooner or later this equipment must be replaced. Thus, the force must be modernized whether or not it is "transformed."

[3]It appears that the current U.S. force structure cannot be maintained within likely future budget levels. Transformation is needed to shift the force to smaller but more capable forces. Otherwise the U.S. military will lose capabilities and the nation's ability to shape the international environment will be reduced. Transformation is therefore a *necessity* for the DoD, not an option. (See Davis et al., 1998.)

We first briefly describe the force transformation effort.

TODAY'S FORCE TRANSFORMATION/RMA ACTIVITIES

The current efforts to "transform the force" are broad based, extending across the DoD. DoD components involved include OSD, the Joint Staff, the Atlantic Command (ACOM), the Services (Army, Navy, Marine Corps, and Air Force), and the Defense Science Board (DSB). Force transformation activities thus far include:[4]

- The development of several *future visions of warfare*, including the Chairman of the Joint Chiefs of Staff's *Joint Vision 2010*, the Air Force's "Global Engagement," the Army's "Army Vision XXI," the Navy's "Forward . . . from the Sea," and the Marine Corps' "Operational Maneuver from the Sea."
- The establishment of a number of *laboratories* dedicated to exploring new ways of warfare, including the Joint Warfighting Center, the Joint Battle Center, a number of Army and Air Force Battle Labs, the Navy Sea-Based Battle Lab, and the Marine Corps Warfighting Lab.
- A number of *wargames* exploring new ways of warfare, including OSD Net Assessment wargames, the Army After Next Wargames, the Navy's Global Wargame series and Strategic Concepts Wargames, the Marine Corps Concept Game series, and the Air Force's Global Engagement Wargames.
- A number of *developmental and field experiments*, including Advanced Concept Technology Demonstrations (ACTDs) and Advanced Technology Demonstrations (ATDs) conducted under the sponsorship and supervision of the Undersecretary of Defense for Acquisition and Technology (USDA&T), Joint experiments sponsored by the Joint Staff (e.g., the J-6's Information Superiority experiments), Army Advanced Warfighting experi-

---

[4]This listing of DoD force transformation-related activities is based on material from Edward L. Warner III (Assistant Secretary of Defense, Strategy and Threat Reduction), "Preparing Now for an Uncertain Future: Modernization and the RMA," and George T. Singley III (Acting Director, Defense Research and Engineering), "DoD Research and Development: Planning for Military Modernization," briefings presented at Defense Week's 18th Annual Defense Conference, December 10, 1997.

ments (e.g., Warrior Focus and Force XXI), Navy Fleet Battle experiments and "Distant Thunder" Antisubmarine Warfare (ASW) experiments, Marine Sea Dragon experiments, and Air Force Expeditionary Force experiments.

- *New organizational arrangements,* including the Army's brigade-sized Experimental Force (EXFOR), and the Air Force's Air Expeditionary Forces and information warfare (IW) and unmanned aerial vehicle (UAV) squadrons.

These activities are pursuing various technology/device/system/operational employment concept combinations, many of which probably represent evolutionary improvements on current ways of waging war, but several of which could possibly lead to RMAs. Among specific concepts proposed as the kernel of "the RMA" are the following:

- *Long-Range Precision Fires.* This RMA candidate was proposed by Andrew Marshall (Director, OSD Net Assessment) in his two initial RMA papers (1993 and 1995). It was also proposed, in more detail, by Walter Morrow (CNO Executive Panel) (1997). The essence of this idea is expressed by Marshall (1995):[5]

  > Long-range precision strike weapons coupled to very effective sensors and command and control systems will come to dominate much of warfare. Rather than closing with an opponent, the major operational mode will be destroying him at a distance.

- *Information Warfare.* This RMA candidate was also proposed by Andrew Marshall in his two RMA papers. It has also been implicitly proposed by Roger Molander and his colleagues (1996) in

---

[5]Long-range precision fires have been under active development for at least 20 years. Morrison and Walker (1978) quote William Perry (then Director of Defense Research and Engineering) as saying:

> [The United States is] converging very rapidly [on three objectives:] to be able to see all high-value targets on the battlefield at any time, to be able to make a direct hit on any target we can see, and to be able to destroy any target we can hit . . . [in order to] make the battlefield untenable for most modern forces.

Although not expressed in RMA terms, this is clearly an early expression of the long-range precision fires concept.

their work on strategic information warfare. The essence of this idea is also expressed by Marshall (1995):

> The information dimension or aspect of warfare may become increasingly central to the outcome of battles and campaigns. Therefore, protecting the effective and continuous operation of one's own information systems, and being able to degrade, destroy or disrupt the functioning of the opponent's, will become a major focus of the operational art.

- *System of Systems.* This RMA candidate was first explicitly proposed by Admiral William A. Owens (Vice Chairman, Joint Chiefs of Staff) (1996). It has subsequently been elaborated on by Blaker (1997). The essence of this idea is that combining a vast assemblage of intelligence collection, surveillance, and reconnaissance (ISR); advanced command, control, communications, computers, and intelligence processing (C^4I); and precision-weapon systems results in a whole with capabilities much greater than the sum of the parts.[6]
- *Network-Centric Warfare.* This RMA candidate was proposed by Vice Admiral Arthur K. Cebrowski and his colleagues in Joint Staff/J-6 (Cebrowski and Garstka, 1998). The network-centric warfare concept employs an operational architecture involving three grids to enable the operational objectives of JV2010: an "Information Grid," a "Sensor Grid," and an "Engagement Grid." The Information Grid provides the computing and communications backbone for the other two grids. The Sensor Grid is an assemblage of space, air, ground, sea, and cyberspace sensors and sensor tasking, processing, and fusion applications, providing battlespace awareness. The Engagement Grid, an assemblage of platforms and weapons, exploits this battlespace awareness to enable the JV2010 force employment objectives of precision engagement, dominant maneuver, and full-dimensional protec-

---

[6]Somewhat earlier, Perry (1991 and 1994) also discussed the system-of-systems concept, although not strictly in the RMA context.

tion. Each of these three grids is connected and functions in a network fashion.[7]

- *Cooperative Engagement Capability.* This concept has been proposed, developed, and demonstrated by the U.S. Navy.[8] The essence of this concept as applied to a Navy battlegroup is that

  > combat systems [on geographically separated platforms] share unfiltered sensor measurement data associated with tracks with rapid timing and precision to enable the battlegroup units to operate as one [in their engagement of enemy targets]. (APL, 1995.)

  Rather than a stand-alone RMA candidate, this concept should probably be thought of as an important harbinger of network-centric warfare.

There are undoubtedly additional items that could be added to this list.

DOES ANYTHING APPEAR TO BE MISSING?

Does anything appear to be missing from these DoD force transformation/RMA activities? Based on the history of past RMAs and the RMA checklist we developed in Chapter Six, the answer seems to be "yes." Table 7.1 summarizes our assessment; we elaborate in what follows.

Enabling Technologies

The ongoing information revolution is clearly providing a fertile set of enabling technologies.

---

[7]The notion of an RMA emerging from networking distributed sensors and weapons had earlier been broached in a never distributed 1991 study, "Project 2025," by the National Defense University's Institute for National Strategic Studies. The technology portions of Project 2025 were later published in Libicki (1994) and in Arquilla and Ronfeldt (1997, Chapter 8).

[8]See APL (1995).

Table 7.1

Does Anything Appear to Be Missing from DoD's Current RMA Activities?

RMA Checklist	DoD's Current Situation
You must have a fertile set of enabling technologies	*We clearly have this*
You must have unmet military challenges	*We have several of these (but are they compelling enough?)*
You must have a receptive organizational climate	*We may have this in some Services (but not in others)*
You must have support from the top	*We have this (but does it include all of the Services?)*
You must have mechanisms for experimentation (to discover, learn, test, and demonstrate)	*We have these (but do they encourage "risky" experiments?)*
You must focus on a definite "thing" or a short list of "things"	*Thus far, this seems to be missing*
You must ultimately challenge someone's core competency	*Thus far, this seems to be missing*
You must have ways of responding positively to successful experiments (in terms of doctrine, acquisition, and force structure)	*This could be a problem (can the DoD system respond positively to a risky new idea?)*

Military Challenges

Even though the United States is now commonly believed to be the world's only superpower, there are still military tasks it cannot perform as confidently as it would like in a wide range of circumstances. Moreover, these unmet (or at least not totally met) challenges are likely to grow in number.[9]

But are these challenges, particularly those future challenges that are not here yet for all to see and none to deny, compelling enough to

[9]The QDR (Cohen, 1997) sketches out some of these challenges; RAND work on asymmetric threats (Bennett et al., 1994a, 1998, 1999) identifies still others.

cause conservative military organizations to accept substantial change? The jury is still out.[10]

Organizational Climate

There appears to be a receptive organizational climate in some of the Services; reports from other Services are mixed.

Support from the Top

We conclude there is support at the very top in the DoD: i.e., the Secretary of Defense, the Chairman of the Joint Chiefs of Staff, and their immediate subordinates. But does this support extend across all branches of all four Services? Reports are mixed.

Mechanisms for Experimentation

As our earlier listing of developmental and field experiments indicates, there are a large number of RMA-related experiments going on in the DoD today. But these experiments may not cover the entire discover, learn, test, and demonstrate spectrum. It appears (at least to this author) that too many of these experiments are "success oriented"; that too many of the experimenters do not feel free to take the kinds of chances necessary to really discover and learn what works and what does not work, what makes sense and what does not; and that too many of these experimental mechanisms do not encourage risky experiments and tolerate failure.

Focus on a Definite "Thing"

DoD's current RMA activities clearly lack focus on one definite "thing" or a short list of "things" which will be the central kernel of "the RMA." Such a focus is still to come. It is not apparent how the

---

[10]Davis et al. (1998) have proposed a set of "operational challenges" for the Secretary of Defense to use as a management technique to motivate and focus the Services' force transformation efforts. Future events will determine the efficacy of such OSD-imposed planning challenges.

vast panoply of DoD RMA-related experiments now under way will bring about such a focus.[11]

Challenging Someone's Core Competency

There is no evidence that anyone anywhere in the DoD is deliberately setting out to challenge a core competency of one of the Services.[12] Until that happens, we will not have an RMA.[13]

Ways of Responding Positively to Successful Experiments

In principle, each of the Services has mechanisms for making doctrinal changes and the DoD has well-established procedures for acquiring new systems and modifying force structures, all of which could respond to successful RMA-related experiments. In principle. But, in practice these formal mechanisms and procedures work best

---

[11]The current lack of focus of DoD's force transformation/RMA activities is reminiscent of the situation in the U.S. Navy's aviation community in the late 1910s and early 1920s. At that time, the Navy was experimenting with many different combinations of air vehicles, basing concepts, and mission applications. It was several years before the Navy began focusing on wheeled planes, based on flat-deck ships, used to attack naval targets—i.e., the essence of carrier aviation and what became the carrier warfare RMA.

[12]In principle, it is immaterial whether a U.S.-led revolution is challenging one of its own core competencies or someone else's. In either case, if the United States succeeds in upsetting that core competence, by the author's definition it will have achieved a revolution in military affairs—it will have overturned the established military order and replaced it with a new order (in some military arena).

A U.S.-led RMA could affect several core competencies other than its own. There are arenas of conflict in which the United States is not superior, not the dominant player. Terrorism and counterterrorism are examples. Today, and for the last few decades, a number of terrorist groups have had a capable core competency to cause substantial civilian and military casualties, and the United States's capabilities to counter, prevent, and defeat such attacks have been limited. Guerrilla warfare, particularly in cities, is another example. The United States did not handle this well during the Vietnam War or in Somalia. A number of additional examples can be found in the general area of asymmetric strategies.

Having said all this, however, the most profound changes in warfare would occur if the United States successfully challenged one of its own core competencies.

[13]Today the core competencies embodied in the tank, manned aircraft, and aircraft carrier appear to be sacred in their respective Services, with no significant in-Service challenges allowed. Where are the challenges (and challengers) in the U.S. military today? None is apparent to this author. Without such challenges and challengers, the United States may miss out on one or more important RMAs.

when the doctrinal changes are small and do not challenge anyone's traditional ways of waging war, the new systems to be acquired represent evolutionary improvements on existing systems, and the force structure modifications are minor, not major. It has been a long time since the formal DoD doctrinal, acquisition, and force-structure-modifying systems have had to respond to radical change.[14] It is unclear how well they will do.[15]

"THE RMA": WHERE WE SEEM TO BE TODAY

Using Secretary Cohen's QDR terminology to describe the force transformation process (in spite of the reservations expressed earlier concerning this choice of words), where is "the RMA" today? Harking back to the model of the RMA process in Figure 3.2, we can say the following:

- *New technology.* We have a lot of this.
- *New devices and systems.* We have a lot of ideas for new devices and systems. Many (but not all) of them have been or are being built. Some (but not most) of them are undergoing experiments, but not necessarily risky experiments covering the entire discover, learn, test, and demonstrate spectrum.
- *New operational concepts.* We have many of these, each with their advocates and detractors. A few are undergoing actual experiments. Most are in paper discussions and arguments.
- *New doctrine and force structure.* We are a long way from this, a very long way.

We are also a long way from focusing on a short list of potentially revolutionary devices, systems, and operational concepts around

[14]In recent years, most truly novel/innovative systems (e.g., the F-117A) have been acquired via "black programs," not through the formal acquisition system.

[15]The recent cancellation of the Arsenal Ship because of lack of Navy support, without building even one to try out the concept, is not reassuring in this regard. Imagine what would have happened to the development of carrier aviation technology and operational concepts in the 1920s and 1930s if the Navy had canceled the 1919–1922 conversion of the collier *Jupiter* into the carrier *Langley* (CV-1), the Navy's first experimental carrier.

which we can "transform the force." This necessary focusing process could take a few years, probably will take several years, and possibly will take many years. Until it happens, we are where the U.S. Navy was in 1920: a long way from an RMA, a long way from being able to transform the force.

Another concern: In most past RMAs, the force wasn't "transformed"—i.e., old force structure elements replaced by RMA elements—until the RMA had been proven in battle. Until then, the RMA elements were treated as add-ons to the then-existing force structure.[16] Based on the QDR, the DoD appears to be planning to "transform the force," i.e., replacing old elements with new RMA elements rather than merely adding those elements, before the RMA is proven in combat. This flies in the face of history.

SOME KEY QUESTIONS FOR THE DoD

We are left with some key questions concerning DoD's current force transformation/RMA activities:

- Can the DoD bring about a true RMA without ultimately challenging one or more of the Services' current core competencies?
- Can the DoD bring about a true RMA without focusing on a definite "thing" or a short list of "things"?
- Can the DoD bring about a true RMA using its current acquisition process?
- Can the DoD "transform the force" to the extent postulated in the QDR (a lot more capability for a lot less resources) without bringing about one or more true RMAs, in the sense defined here?

The author fears the answer to each of these questions is "no."

WHAT CAN BE DONE TO FILL IN THE MISSING ELEMENTS?

Three things in particular seem to be needed:

[16]In the 1930s, the German army added panzer divisions, but it kept all of its infantry divisions; the U.S. Navy added aircraft carriers, but it kept all of its battleships.

- Some mechanism to encourage challenges (over time) to one or more of the Services' current core competencies
- Some mechanism to bring about a focus (over time) of the force transformation/RMA process on a definite "thing" or a (fairly) short list of "things"
- Changes in the DoD acquisition system to make it more receptive to novel/radical innovations.

Meeting the first of these needs, *encouraging challenges to core competencies*, initially requires one or more multidisciplinary groups of creative people, with technology, military systems, and military operations backgrounds, working together for an extended period to conceptualize new systems and operational concepts that challenge one or more core competencies. These concept groups must be free to challenge whatever Service core competency their expertise, vision, and intuition tell them is ripe to be overturned, with nothing held sacred.[17]

Continuing the RMA process beyond the conceptual stage and meeting the second of the above mentioned needs—*bringing about a focus of the force transformation process on a definite "thing"*—requires one or more *experimental groups* that will lay out experimental

---

[17]Others have proposed entities similar to our concept groups. Krepinevich (1995) proposes a permanent Concept Development Center (CDC) to "facilitate the 'intellectual breakthroughs' in operational concepts, and in corresponding military systems and organizations . . . to provide the foundation for successful U.S. [military] adaptation." He estimates that such an organization should have roughly 100 individuals.

In a similar vein, the CNO Executive Panel's Naval Warfare Innovation Task Force proposes the establishment of Concept Generation Team(s) to accomplish much the same purpose, limited, however, to naval warfare. These teams would have the following characteristics: "CNO to enunciate objectives; strong leadership (RADM level) providing continuity over several years; location at naval education facility(s) well insulated from Washington, D.C.; separate activity from current function of facility; small team(s) (<10) of Navy and Marine Corps' most innovative thinkers at O-5/O-6 levels and equivalent level civilians; ability to utilize most innovative members of current faculties and student classes; ability to draw on resources of leading universities; teams to operate on a 4–6 month temporary duty basis with outputs briefed to CNO; outputs analyzed promptly by independent resident capability." (Harris, briefing to RAND, 1998).

Alternatively, the RAND Concept Options Group (COG) construct might be used as the foundation for a number of temporary concept groups (see Birkler et al., 1998).

roadmaps (covering the entire discover, learn, test, and demonstrate spectrum) for promising concepts created by the concept groups, and then design, conduct, and evaluate the experiments, iterating the process as many times as required, for as long as is required (with no artificial deadlines).[18] These experimental groups must be free to take risks and to fail (from time to time). They should interact frequently with the concept groups; they could be part of the same organization(s).

Establishing the proper organizational position(s) for both the concept and experimental groups relative to the mainstream military is complicated and problematic. On the one hand, the CNO Executive Panel emphasizes the importance of separating such conceptual and experimental activities involving revolutionary innovation from the mainstream activity of the military organization(s), so that they will be free to take chances and truly challenge core competencies.[19] On the other hand, Murray and Watts (1995) emphasize the danger of such innovative activities being too separated from the mainstream military organization(s), in which case they may be viewed as outsiders and their new ideas may not be accepted by the people who will fight the next war.

Thus, the concept groups and experimental groups should be somewhat separated from the mainstream military, but not too much—clearly a delicate task in organizational design. One possible solution, suggested by Birkler et al. (1999), is to create *provisional operational units* that would participate in the (learn, test, and demonstrate) experiments along with the experimental groups, thereby accumulating field operating experience, developing military user "buy in" for the new systems, motivating and informing the necessary doctrinal developments, and (if all of this is successful) providing an early, limited but useful combat capability.[20]

Meeting the third of the above mentioned needs—*making the DoD acquisition system more receptive to novel/radical innovations*—

[18]It took the Bureau of Aeronautics about ten years to establish the foundations of carrier aviation.

[19]See footnote 10 in Chapter Five.

[20]Such units have been tried in the past. The initial F-117A squadrons were of this nature, as was the USS *Langley* (CV-1).

requires that a new branch be added to the acquisition system.[21] As discussed in Chapter Six, this "RMA branch" of the acquisition system should tolerate substantial military-utility risks to a much later stage in the acquisition process, in some cases all the way to IOC. The phases and decision points in this new acquisition branch should be constructed in keeping with the likely uncertainties at each stage in the RMA process. Figure 7.1 and Table 7.2 illustrate what this new branch might look like in terms of acquisition phases and milestones.[22]

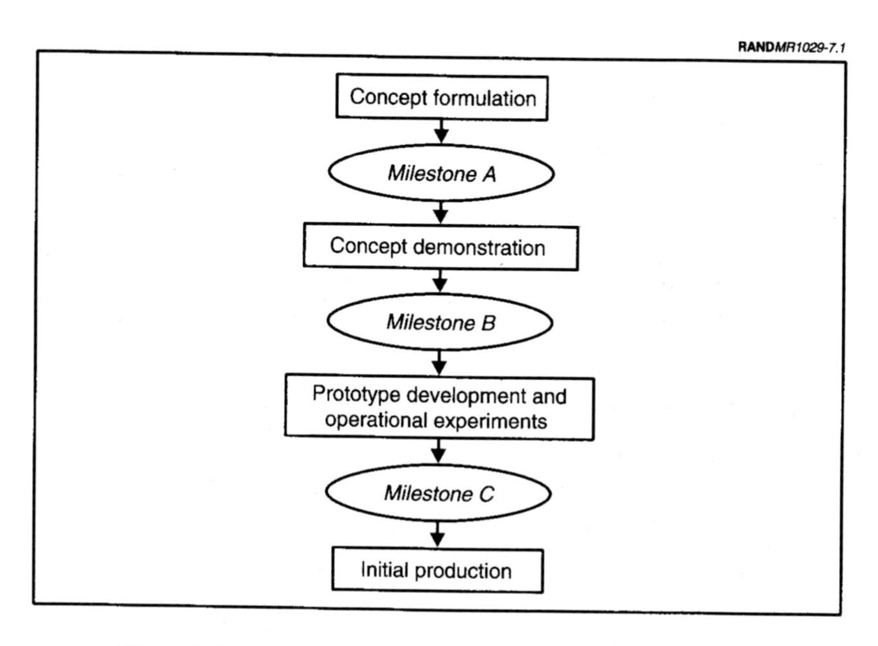

Figure 7.1—An RMA Branch of the DoD Acquisition System

---

[21]John Birkler and his RAND colleagues have proposed a similar branch be added to the DoD acquisition process to handle novel and radical innovations, whether or not they qualify as RMAs.

[22]The acquisition phases and milestones shown in Figure 7.1 and Table 7.2 are similar to but slightly different from those proposed by Birkler et al. (1999).

Table 7.2

A Possible Set of Milestones for an "RMA Branch" of the DoD Acquisition System

Milestone	Certainties Required	Uncertainties and Risks Tolerated
Milestone A	If it works as advertised, it could bring about a major increase in U.S. military capabilities.	Technical feasibility, exact operational system configuration, exact operational employment concept, differences of opinion regarding military utility.
Milestone B	Proof-of-concept demonstration of major technical issues.	Exact operational system configuration, exact operational employment concept, differences of opinion regarding military utility.
Milestone C	Operational system configuration determined; initial operational employment concept established; plausible case made for military utility.	Final operational employment concept; some differences of opinion regarding military utility.

Doing these four things—setting up concept groups, experimental groups, and provisional operational units, and adding a new branch to the DoD acquisition system—should go a long way toward filling in the missing elements in DoD's current force transformation/RMA activities.

IN SUMMARY

Based on the history of past RMAs, there appear to be missing elements in DoD's current force transformation activities:

- None of the Services' current core competencies are being challenged.
- There is inadequate focus on a definite "thing" or a short list of "things."
- The DoD acquisition system may not be adequately receptive to novel/radical innovations.

These missing elements can be filled by:

- Setting up DoD concept and experimental groups to identify and experiment with new systems and operational concepts that (a) challenge current Service core competencies and (b) increase the focus of the current RMA efforts.
- Establishing provisional operational units to participate in experiments with new systems and operational concepts.
- Establishing a new branch to the DoD acquisition system that tolerates military-utility risks to a much later stage in the acquisition process.

Doing the above will facilitate DoD's force transformation activities and help ensure that the next RMA is brought about by the United States and not some other nation.

BIBLIOGRAPHY

(APL, 1995) "The Cooperative Engagement Capability," *John Hopkins APL Technical Digest,* Vol. 16, No. 4, 1995, pp. 377–396.

(APS, 1987) *Science and Technology of Directed Energy Weapons,* Report of the American Physical Society Study Group, The American Physical Society, April 1987.

(Arquilla and Ronfeldt, 1997) John Arquilla and David F. Ronfeldt (eds.), *In Athena's Camp: Preparing for Conflict in the Information Age,* RAND, MR-880-OSD/RC, 1997.

(Aviation Week, 1996) David A. Fulghum, "Boeing Team Tapped To Build Laser Aircraft," *Aviation Week & Space Technology,* November 18, 1996, pp. 22–23.

(Barker, 1992) Joel Arthur Barker, *Paradigms, The Business of Discovering the Future,* William Morrow and Company, Inc., New York, 1992.

(Barnett, 1996) Jeffrey R. Barnett, *Future War: An Assessment of Aerospace Campaigns in 2010,* Air University Press, Maxwell AFB, Alabama, 1996.

(Bartlett et al., 1996) Henry C. Bartlett, G. Paul Holman, Jr., and Timothy E. Somes, "Force Planning, Military Revolutions and the Tyranny of Technology," *Strategic Review,* Vol. 24, Fall 1996, pp. 28–40.

(Bennett et al., 1994a) Bruce W. Bennett, Sam Gardiner, Daniel B. Fox, and Nicholas K. J. Witney, *Theater Analysis and Modeling in an Era of Uncertainty: The Present and Future of Warfare*, RAND, MR-380-NA, 1994.

(Bennett et al., 1994b) Bruce W. Bennett, Sam Gardiner, and Daniel B. Fox, "Not Merely Planning for the Last War," in Paul K. Davis (ed.), *New Challenges for Defense Planning*, RAND, MR-400-RC, 1994, pp. 477–514.

(Bennett et al., 1998) Bruce W. Bennett, Christopher P. Twomey, and Gregory F. Treverton, *Future Warfare Scenarios and Asymmetric Threats*, RAND, MR-1025-OSD, 1998.

(Bennett et al., 1999) Bruce W. Bennett, Christopher P. Twomey, and Gregory F. Treverton, *What Are Asymmetric Strategies?* RAND, DB-246-OSD, 1999.

(Birkler et al., 1998) John Birkler, C. Richard Neu, and Glenn Kent, *Gaining New Military Capability: An Experiment in Concept Development*, RAND, MR-912-OSD, 1998.

(Birkler et al., 1999) John Birkler, Giles Smith, Glenn Kent, and Robert Johnson, *Streamlining the Process of Modernizing Our Forces*, RAND, forthcoming.

(Blaker, 1997) James R. Blaker, *Understanding the Revolution in Military Affairs: a Guide to America's 21st Century Defense*, Defense Working Paper No. 3, Progressive Policy Institute, 518 C Street, NE, Washington, DC 20002, 1997.

(Bower and Christensen, 1995) Joseph L. Bower and Clayton M. Christensen, "Disruptive Technologies: Catching the Wave," *Harvard Business Review*, January–February 1995, pp. 43–53.

(Bowie et al., 1993) Christopher Bowie, Fred Frostic, Kevin Lewis, John Lund, David Ochmanek, and Philip Propper, *The New Calculus, Analyzing Airpower's Changing Role in Joint Theater Campaigns*, RAND, MR-149-AF, 1993.

(Brodie, 1973) Bernard and Fawn Brodie, *From Crossbow to H-Bomb, The Evolution of the Weapons and Tactics of Warfare*, Indiana University Press, Bloomington, Indiana, 1973.

(Buchan, 1998) Glenn C. Buchan, *One-and-One-Half Cheers for the Revolution in Military Affairs*, RAND, P-8015, 1998.

(Burke, 1978) James Burke, *Connections*, Little, Brown and Company, Boston, 1978.

(Cebrowski and Garstka, 1998) Arthur K. Cebrowski and John J. Garstka, "Network-Centric Warfare: Its Origin and Future," *U.S. Naval Institute Proceedings*, January 1998, pp. 28–35.

(Cerf and Navasky, 1984) Christopher Cerf and Victor Navasky, *The Experts Speak*, Pantheon Books, New York, 1984.

(Christensen, 1997) Clayton M. Christensen, *The Innovator's Dilemma: When New Technologies Cause Great Firms to Fail*, Harvard Business School Press, Cambridge, Massachusetts, 1997.

(Churchill, 1948) Winston S. Churchill, *The Second World War*, Vol. 1, *The Gathering Storm*, Houghton, Mifflin Company, Boston, 1948.

(Churchill, 1949) Winston S. Churchill, *The Second World War*, Vol. 2, *Their Finest Hour*, Houghton, Mifflin Company, Boston, 1949.

(Churchill, 1958) Winston S. Churchill, *The History of the English Speaking Peoples*, Vol. I, *The Birth of Britain*, Dodd, Mead & Company, New York, 1958.

(Cohen, 1997) William S. Cohen (Secretary of Defense), *Report of the Quadrennial Defense Review*, Office of the Secretary of Defense, May 1997.

(Corum, 1992) James S. Corum, *The Roots of Blitzkrieg: Hans von Seeckt and German Military Reform*, University Press of Kansas, Lawrence, Kansas, 1992.

(Davis et al., 1998) Paul K. Davis, David C. Gompert, Richard J. Hillestad, and Stuart Johnson, *Transforming the Force: Suggestions for DoD Strategy*, RAND, IP-179, 1998.

(Dewar, 1993) James A. Dewar, *Assumption-Based Planning: A Planning Tool for Very Uncertain Times*, RAND, MR-114-A, 1993.

(DoD, 1996) *Defense Acquisition,* Department of Defense Directive 5000.1, March 15, 1996.

(DoD, 1997) *Mandatory Procedures for MDAPs and MAIS Acquisition Programs,* Department of Defense 5000.2-R, October 6, 1997.

(Doughty, 1985) Robert Allan Doughty, *The Seeds of Disaster: The Development of French Army Doctrine, 1919–1939,* Archon Books, Hamden, Connecticut, 1985.

(Doughty, 1990) Robert Allan Doughty, *The Breaking Point: Sedan and the Fall of France, 1940,* Archon Books, Hamden, Connecticut, 1990.

(DSB, 1990) *Executive Summary of Defense Science Board 1990 Summer Study on Research and Development Strategy for 1990s,* Defense Science Board, Department of Defense, August 8, 1990.

(DSB, 1996) *Tactics and Technology for 21st Century Military Superiority,* Defense Science Board, Department of Defense, 1996.

(Dupuy, 1966) Trevor N. Dupuy, *Strategic Concepts and the Changing Nature of Modern War,* Historical Evaluation and Research Organization, Vol. 1, Washington, DC, 1966, p. 239.

(Dupuy, 1984) Trevor N. Dupuy, *The Evolution of Weapons and Warfare,* Da Capo Press, New York, 1984.

(Ellis, 1975) John Ellis, *The Social History of the Machine Gun,* The Johns Hopkins University Press, Baltimore, Maryland, 1975.

(FitzGerald, 1987) Mary C. FitzGerald, *Marshal Ogarkov and the New Revolution in Soviet Military Affairs,* Center for Naval Analyses, CRM 87-2, January 1987.

(Friedman, 1988) Norman Friedman, *British Carrier Aviation: The Evolution of the Ships and Their Aircraft,* Annapolis, Maryland, 1988.

(Gray, 1995) Colin S. Gray, "The Changing Nature of Warfare," *Naval College Review,* Vol. XLIX, No. 2, 1995, pp. 7–22.

(Grove, 1996) Andrew S. Grove, *Only the Paranoid Survive, How to Exploit the Crisis Points that Challenge Every Company and Career*, Doubleday, New York, 1996.

(Guderian, 1952) Heinz Guderian, *Panzar Leader*, translated from the German by Constantine Fitzgibbon, E. P. Dutton & Co., New York, 1952.

(Herzfeld, 1991) Charles M. Herzfeld, Director, Defense Research and Engineering, *Statement on Defense Technology*, to the Subcommittee on Research and Development of the Committee on Armed Services, United States House of Representatives, April 23, 1991.

(JDR, 1986) "High Energy Lasers: Systems Concepts and Technology," *Journal of Defense Research*, Special Issue 86-1, May 1986.

(Johnson, 1990) David E. Johnson, *Fast Tanks and Heavy Bombers: The United States Army and the Development of Armor and Aviation Doctrines and Technologies, 1917–1945*, PhD Thesis, Duke University, Durham, North Carolina, 1990.

(Johnson, 1998) David E. Johnson, *Fast Tanks and Heavy Bombers: Innovation in the U.S. Army, 1917–1945*, Cornell University Press, Ithaca, New York, 1998.

(Joint Staff, 1997) *Concept for Future Joint Operations: Expanding Joint Vision 2010*, Joint Chiefs of Staff, Washington, DC, 1997.

(Kendall, 1992) Frank Kendall, "Exploiting the Military Technical Revolution: A Concept for Joint Warfare," *Strategic Review*, Spring 1992, pp. 23–30.

(Krepinevich, 1994) Andrews F. Krepinevich, "Cavalry to Computer: The Pattern of Military Revolution," *The National Interest*, Fall 1994, pp. 30–42.

(Krepinevich, 1995) Andrews F. Krepinevich, *The Military Evolution: Restructuring Defense for the 21st Century*, statement prepared for the Subcommittee on Acquisition & Technology, Senate Armed Services Committee, May 5, 1995.

(Kuhn, 1970) Thomas S. Kuhn, *The Structure of Scientific Revolutions,* University of Chicago Press, Illinois, 1970.

(Libicki, 1994) Martin Libicki, *The Mesh and the Net: Speculations on Armed Conflict in an Age of Free Silicon,* McNair Paper 28, National Defense University, Fort McNair, Washington, DC, March 1994.

(Libicki, 1996) Martin Libicki, *Information & Nuclear RMAs Compared,* Strategic Forum Number 82, Institute of National Strategic Studies, National Defense University, Fort McNair, Washington, DC, June 1996.

(Libicki, 1999) Martin C. Libicki, *Illuminating Tomorrow's War,* McNair Paper 60, Institute of National Strategic Studies, National Defense University, Fort McNair, Washington, DC, 1999.

(Libicki and Hazlett, 1994) Martin Libicki and James Hazlett, *The Revolution in Military Affairs,* Strategic Forum Number 11, Institute of National Strategic Studies, National Defense University, Fort McNair, Washington, DC, 1994.

(Liddell Hart, 1979) Basil H. Liddell Hart (ed.), *The German Generals Talk,* Quill, New York, 1979.

(Macksey, 1975) Kenneth Macksey, *Guderian, Creator of the Blitzkrieg,* Stein and Day, New York, 1975.

(Mann, 1998) Paul Mann, "Revisionists Junk Defense Revolution," *Aviation Week & Space Technology,* April 27, 1998, pp. 37–38.

(Marshall, 1993) Andrew W. Marshall, *Some Thoughts on Military Revolutions,* Memorandum for the Record, OSD Office of Net Assessment, July 27, 1993.

(Marshall, 1995) Andrew W. Marshall, Director of Net Assessment, OSD, *Revolutions in Military Affairs,* statement prepared for the Subcommittee on Acquisition & Technology, Senate Armed Services Committee, May 5, 1995.

(Mazarr, 1994) Michael J. Mazarr, *The Revolution in Military Affairs: A Framework for Defense Planning,* Strategic Studies Institute, U.S. Army War College, Carlisle Barracks, Pennsylvania, June 10, 1994.

(Mazarr et al., 1993) Michael J. Mazarr, Jeffrey Shaffer, and Benjamin Ederington, *The Military Revolution, A Structural Framework,* Center for Strategic and International Studies, 1800 K Street, NW, Washington, DC, March 1993.

(McGrath and MacMillan, 1995) Rita G. McGrath and Ian C. MacMillan, "Discovery-Driven Planning," *Harvard Business Review,* July–August 1995, pp. 4–12.

(McKendree, 1996) Tom McKendree, *The Revolution in Military Affairs—Issues, Trends, and Questions for the Future,* paper presented at 64th MORS Conference, Fort Leavenworth, Kansas, June 1996.

(Melhorn, 1974) Charles M. Melhorn, *Two-Block Fox: The Rise of the Aircraft Carrier, 1911–1929,* Naval Institute Press, Annapolis, Maryland, 1974.

(Mitchell, 1921) William Mitchell, *Our Air Force: The Keystone to National Defense,* E. P. Dutton & Co., New York, 1921.

(Mitchell, 1925) William Mitchell, *Winged Defense: The Development and Possibilities of Modern Air Power—Economic and Military,* G. P. Putnam's Sons, New York, 1925.

(Molander et al., 1996) Roger Molander, Andrew Riddile, and Peter Wilson, *Strategic Information Warfare,* RAND, MR-661-OSD, 1996.

(Morison, 1963) Samuel Elliot Morison, *The Two-Ocean War, A Short History of the United States Navy in the Second World War,* Little, Brown and Company, Boston, 1963.

(Morrison and Walker, 1978) Philip Morrison and Paul F. Walker, "A New Strategy for Military Spending," *Scientific American,* Vol. 239, No. 4, October 1978, pp. 48–61.

(Morrow, 1997) Walter E. Morrow, Jr., "Technology for a Naval Revolution in Military Affairs," presentation to CEP-RMA Round Table, June 4, 1997.

(Murray and Millet, 1996) Williamson Murray and Allan R. Millet (eds.), *Military Innovation in the Interwar Period,* Cambridge University Press, Cambridge, Massachusetts, 1996.

(Murray and Watts, 1995) Williamson Murray and Barry Watts, *Military Innovation in Peacetime*, report prepared for OSD Net Assessment, January 20, 1995.

(NSB, 1997) *Technology for the United States Navy and Marine Corps, 2000–2035*, Naval Studies Board, National Research Council, National Academy Press, Washington, DC, 1997.

(Owens, 1996) William A. Owens, "The Emerging U.S. System-of-Systems," *Strategic Forum*, No. 63, Institute of National Strategic Studies, National Defense University, Fort McNair, Washington, DC, 1996.

(Perry, 1991) William J. Perry, "Desert Storm and Deterrence," *Foreign Affairs*, Vol. 70, Fall 1991, pp. 66–82.

(Perry, 1994) William J. Perry, "Military Action: When to Use It and How to Ensure Its Effectiveness," in Janne E. Nolan (ed.), *Global Engagement: Cooperation and Security in the 21st Century*, The Brookings Institution, Washington, DC, 1994, p. 240.

(Rhodes, 1986) Richard Rhodes, *The Making of the Atomic Bomb*, Simon & Schuster, New York, 1986.

(Ricks, 1994) Thomas S. Ricks, "Warning Shot: How Wars Are Fought Will Change Radically, Pentagon Planner Says," *The Wall Street Journal*, July 15, 1994, p. A1.

(Rosen, 1991) Stephen Peter Rosen, *Winning the Next War: Innovation and the Modern Military*, Cornell University Press, Ithaca, New York, 1991.

(SAB, 1995) *New World Vistas: Air and Space Power for the 21st Century*, USAF Scientific Advisory Board, 1995.

(Schwartz, 1991) Peter Schwartz, *The Art of the Long View*, Doubleday/Currency, New York, 1991.

(Senge, 1990) Peter M. Senge, *The Fifth Discipline: The Art and Practice of the Learning Organization*, Doubleday/Currency, New York, 1990.

(Shalikashvili, 1996) John M. Shalikashvili (Chairman of the Joint Chiefs of Staff), *Joint Vision 2010*, Office of the Chairman of the Joint Chiefs of Staff, July 1996.

(Toffler, 1993) Heidi and Alvin Toffler, *War and Anti-War: Survival at the Dawn of the Twenty-First Century*, Little, Brown and Company, Boston, 1993.

(Turnbull and Lord, 1949) Archibald D. Turnbull and Clifford L. Lord, *History of United States Naval Aviation*, Yale University Press, New Haven, Connecticut, 1949.

(Utterback, 1994) James M. Utterback, *Mastering the Dynamics of Innovation, How Companies Can Seize Opportunities in the Face of Technological Change*, Harvard Business School Press, Cambridge, Massachusetts, 1994.

(van Creveld, 1989) Martin van Creveld, *Technology and War, From 2000 B.C. to the Present*, Collier Macmillan Publishers, London, 1989.

(Watts and Murray, 1996) Barry Watts and Williamson Murray, "Military Innovation in Peacetime," Chapter 10 in Williamson Murray and Allan R. Millet (eds.), *Military Innovation in the Interwar Period*, Cambridge University Press, Cambridge, Massachusetts, 1996, pp. 369–415.

(Weisbord and Janoff, 1995) Marvin R. Weisbord and Sandra Janoff, *Future Search: An Action Guide to Finding Common Ground for Action in Organizations and Communities*, Berrett-Koehler, San Francisco, California, 1995.

(Wilson, 1950) Eugene E. Wilson, "The Navy's First Carrier Task Force," *U.S. Naval Institute Proceedings*, February 1950, pp. 139–169.